Wm. Turner

Lectures on the Comparative Anatomy of the Placenta

First Series

Wm. Turner

Lectures on the Comparative Anatomy of the Placenta
First Series

ISBN/EAN: 9783337162894

Printed in Europe, USA, Canada, Australia, Japan

Cover: Foto ©berggeist007 / pixelio.de

More available books at **www.hansebooks.com**

LECTURES

ON THE

COMPARATIVE ANATOMY

OF THE

PLACENTA.

FIRST SERIES.

DELIVERED BEFORE THE ROYAL COLLEGE OF SURGEONS
OF ENGLAND, JUNE, 1875.

BY

WM. TURNER, M.B. (Lond.)

PROFESSOR OF ANATOMY, UNIVERSITY OF EDINBURGH; LECTURER ON ANATOMY
AND PHYSIOLOGY, ROYAL COLLEGE OF SURGEONS, ENGLAND;
MEMBER OF THE GENERAL MEDICAL COUNCIL.

EDINBURGH:
ADAM AND CHARLES BLACK.
1876.

Cambridge:
PRINTED BY C. J. CLAY, M.A.
AT THE UNIVERSITY PRESS.

PREFACE.

THESE Lectures on the Comparative Anatomy of the Placenta were delivered in the Theatre of the Royal College of Surgeons of England, on the 14th, 15th and 16th June, 1875. They were illustrated by numerous preparations both macroscopic and microscopic, and by many enlarged coloured diagrams. The diagrams were enlarged from the preparations by my assistant Mr A. H. Young. The figures from the microscopic preparations, which are here reproduced on wood, are from drawings kindly made by that gentleman; whilst the figures in the lithographic plates are, with two exceptions, from drawings executed by my former assistant Mr J. C. Ewart, M.B.. to illustrate my Memoir on the Placentation of the Seals in the Transactions of the Royal Society of Edinburgh.

The observations on the Placentæ of the Narwhal, the Lemurs, the Hyrax, and the Elephant, on the chorion of the Giraffe, and on the post partum condition of the mucous membrane of the Cat's uterus, have been made since the Lectures were delivered.

In a second series of Lectures to be given in the summer of 1876, it is intended to describe the Anatomy of this Organ in the other orders of Mammals.

December, 1875.

CONTENTS.

THE COMPARATIVE ANATOMY

OF

THE PLACENTA.

MR PRESIDENT, FELLOWS OF THE COLLEGE AND GENTLEMEN.

WHEN the Council of the College elected me to fill the office of Lecturer on Anatomy and Physiology, I thought that the best mode in which I could show my sense of the honour conferred upon me was to select for consideration and discussion in the Lectures, a department of Anatomy which I had myself investigated. As the Anatomy of the Placenta in the several orders of placental mammals has engaged my attention for some years, and as during that period I have had the good fortune to acquire the gravid uteri of several rare mammals, the placentation of which had been little studied, I have chosen as the subject of my Lectures the Comparative Anatomy of the Placenta; an organ, which, from its importance as the medium of connection between mother and fœtus during intra-uterine life, is worthy of careful study by the Anatomist and Physiologist. In the course of my remarks I hope to submit to your notice several new facts and relations of structure, and to throw some additional light upon previously recorded observations.

It was my original intention to have passed under review during the present course the several modifications in the form and structure of the placenta exhibited by the various orders of placental mammals, and to have concluded with a description of the human placenta, but I found, when I began to arrange the material I had gathered together, that it would have been impossible to compress so extensive a subject into three Lectures, without omitting much that I deemed it important to communicate, both as regards facts in detail and the principles to be deduced from them. I have therefore limited myself on this occasion to an enquiry into the development, form, and structure of the fœtal membranes, the structure of the unimpregnated uterine mucous membrane, and the changes taking place, on the one hand in the chorion, on the other hand in the uterine mucosa, which lead to the formation of the diffused, the polycotyledonary and the zonary forms of placenta. In the course of my description I do not intend to enter at any length into the consideration of the shape and more obvious arrangements of the placenta and membranes, as these have already been fully described by many able anatomists; but to speak more especially of the minute or microscopic structure of the organ, a branch of enquiry of equal, if not of greater importance, than the determination of the macroscopic characters, for only after it has been determined in the several orders of mammals can a correct conception of the general morphology and physiology of the organ be obtained.

In all inquiries into the anatomy of the Placenta, two series of structures have to be investigated, the one belonging to the fœtus, the Fœtal Placenta; the other to the mother, the Maternal Placenta. The placenta is therefore a compound organ, and the complexity of its structure in any given mammal is proportional to the degree in which, in the course of its development and growth, the originally separable fœtal and maternal portions have become interlaced with each other.

THE observations on the structure and development of the ovary, which have been made during the last few years, more especially by Professors Pflüger[1] and Waldeyer[2] and by Dr Foulis[3], have satisfactorily demonstrated that the ova are derived from the corpuscles of the epithelial layer which invests the surface of the ovary. By a process of involution accompanied by a great growth of the ovarian stroma, many of these corpuscles become imbedded in the substance of the ovary. The nucleus of a corpuscle swells out into a spherical germinal vesicle, and contains in its interior a nucleolus, or germinal spot. Around the nucleus the protoplasm of the cell-substance increases greatly in amount to form the yelk. The peripheral portion of this yelk-protoplasm then differentiates into a cell-wall—the zona pellucida, or vitelline membrane—which completes the development of the ovarian ovum. The vascular connective-tissue stroma of the ovary forms a capsule around each ovum, which constitutes the vascular capsule of the Graafian follicle. This capsule is lined at first by a single, but afterwards by more than one layer of cells, which form the cells of the membrana granulosa, or so-called epithelium of the Graafian follicle. These cells were believed by Waldeyer to be derived from certain of the epithelial cells investing the ovary, which had been included in the ovarian substance along with those which developed into ova. But Dr Foulis has pointed out that they are descended from the connective-tissue corpuscles of the vascular stroma. The Graafian follicle enlarges by the multiplication of the cells of the membrana granulosa, and by the secretion of follicular fluid. In the course of time it bursts, and the ovum is extruded. Being received by the Fallopian tube it is conveyed into the cavity of the uterus.

Should the ovum become fertilized by the penetration of spermatozoa through the zona pellucida, a remarkable series of

---

[1] *Die Eierstöcke der Säugethiere und des Menschen.* Leipzig, 1863.
[2] *Eierstock und Ei.* Leipzig, 1870.
[3] Abstract in *Proc. Roy. Soc. Edinburgh,* 1874–75, and *in extenso* in *Transactions* of the same *Society,* 1875.

developmental changes takes place in it, which leads to the production not only of the embryo, but of the fœtal membranes. The germinal vesicle disappears, the yelk undergoes cleavage, and in course of time becomes subdivided into multitudes of minute cells, each consisting of a nucleated clump of protoplasm. These cells are arranged immediately within the zona pellucida to form a cellular membrane, the blastoderm, which in course of time surrounds the undifferentiated central part of the yelk, assumes a vesicular form, and is known to embryologists as the blastodermic vesicle. At one part of the blastoderm a roundish white spot then appears and forms the area germinitiva, or area of formation of the future embryo. Commencing at the area germinitiva the blastoderm splits into two layers, an outer, epiderm or epiblast, and an inner, hypoderm or hypoblast. Subsequently a third or intermediate layer is formed, the mesoderm or mesoblast. By most embryologists the mesoblast is regarded as derived from the hypoblast, but Kölliker, in a recent memoir[1], considers that it has no genetic relation with the hypoblast, but arises from the epiblast by an increase of the cells of the same.

Whilst these changes are taking place in the cells derived from the differentiation of the yelk, the zona pellucida assumes a different appearance. At first it is quite smooth, but soon, at least in the dog and rabbit, numerous short, very fine, simple villi project from its outer surface, and form the villi of the primitive chorion. Whether or not similar villous outgrowths project from the zona pellucida of the human ovum has not yet been determined. The villi are, like the zona itself, perfectly structureless, and their apparent object is to attach the ovum in the early stage of gestation to the surface of the uterine mucous membrane. In a short time the primitive chorion disappears, and is replaced by the secondary or permanent chorion, which forms the outer envelope of the embryo, and is the proper medium of attachment to the wall of the uterus.

The mode of origin of the secondary or persistent chorion has now to be considered. To form a definite conception of its production it will be necessary to glance briefly at the changes

---

[1] *Verh. der Phys. Med. Gesellschaft, Würzburg.* January, 1875.

which take place in the area germinitiva. From the three layers of cells, into which the blastoderm divides in this area, the several tissues and organs of the embryo arise, by a process of multiplication and histological differentiation. Speaking generally, one may say that the epiblast cells give origin to the cuticle and its appendages and to the central organs of the cerebro-spinal nervous system ; the mesoblast cells to the connective, muscular, osseous and vascular systems, and to the distributory part of the nervous system ; whilst from the hypoblast cells the epithelial lining of the alimentary canal and of the various glands which open into it arises. The rapid multiplication of these cells in the area germinitiva leads to the production of the body of the embryo. In the course of time a longitudinal axial depression appears in the epiblast, which marks the position of the brain and spinal cord, and below this groove an axial cord, the chorda dorsalis, arises in the mesoblast, which indicates the position of the future spinal column. On each side of the chorda the mesoblast splits into two layers, which become separated from each other by a space, the future pleuroperitoneal cavity. The outer layer of the mesoblast, or somatopleure, is intimately associated with the epiblast, and forms with it the walls of the body of the embryo ; whilst the inner layer of the mesoblast, or splanchno-pleure, is intimately associated with the hypoblast, and forms with it the walls of the alimentary canal and its appendages.

From the margin of the area germinitiva a delicate fold arises, which, by becoming more and more elevated, gradually extends itself above the back of the young embryo, until at last the folds from opposite sides meet in the middle line, become continuous with each other, and form the membrane of the amnion. Between the amnion and the back of the embryo is a space, the cavity of the amnion, which increases in size by the secretion of the liquor amnii into it. The amnion consists of two layers, an outer and an inner, and as it is derived from the area germinitiva it is necessarily continuous with the blastoderm. The inner layer of the amnion, which lies next the amniotic cavity, consists of tessellated epithelial cells, is derived from the epiblast, and is continuous with the cuticular layer of the embryo. The outer layer of the amnion is a thin stratum, continuous with the somato-pleure layer of the mesoblast, and formed of stellate and spindle-shaped cells, like the corpuscles of embryonic connective tissue[1]. The amnion is the most internal of the fœtal membranes, and forms an ovoid bag in which the fœtus and liquor amnii are contained.

But before the amniotic folds meet to become continuous with each other, the free edge of each fold is bent outwards, and, by a continuous process of growth, gradually spreads around the ovum immediately within the zona pellucida so as to form a layer, which was named by von Baer[2] the serous envelope of the ovum, but which may more appropriately, from its position within the zona, be named the sub-zonal membrane. When the union of the amniotic folds has taken place the sub-zonal membrane becomes completely severed from the proper amnion, and constitutes one of the layers of the secondary or persistent chorion. The sub-zonal membrane consists essentially of a layer of cells, which was originally continuous with the cellular layer lining the inner surface of the proper amnion, and through it with the epiblast cells forming the cuticular covering of the embryo. But it is not unlikely that along with this cell-layer a thin prolongation of the layer of stellate connective-tissue corpuscles of the somato-pleure layer of the mesoblast is present. Between the sub-zonal membrane and the proper amnion a space exists, continuous with the general pleuro-peritoneal cavity of the embryo, which is of interest in connection with the formation of the permanent chorion, as the space into which the allantois grows and expands.

Whilst the changes in the epiblast and somato-pleure, which lead to the formation of the amnion and sub-zonal outer layer of the persistent chorion are taking place, the hypoblast and splanchno-pleure undergo important modifications. They gradually extend over the undifferentiated part of the yelk, which they enclose in a sac, the yelk-sac, or umbilical vesicle, and at the same time they form along the ventral surface of the embryo an elongated tube, the alimentary canal. This canal at first freely communicates with the yelk-sac, but as the walls of the alimentary canal thicken and become more closed in,

---

[1] Schenk, *Lehrbuch der vergleichenden Embryologie.* Wien, 1874.
[2] *Ueber Entwicklungsgeschichte der Thiere.* 1828, p. 79.

the communication narrows, until at length nothing is left but a slender duct, the omphalo-mesenteric duct, as the channel of communication between the umbilical vesicle and the intestinal part of the alimentary tube. In the course of time this duct shrivels up and disappears.

In the splanchno-pleure forming the wall of the umbilical vesicle and in the corresponding wall of the alimentary tube blood-vessels are developed. As that pole of the umbilical vesicle, which lies opposite the omphalo-mesenteric duct reaches, and is, at an early period of its development, in contact with a limited area of the sub-zonal membrane, the blood-vessels in the wall of the vesicle may be, and in some mammals are, conveyed by it up to the sub-zonal membrane.

But an important change also takes place at the pelvic end of the alimentary tube. A small vesicular outgrowth appears, which communicates with the intestine and forms the allantois. The allantois rapidly increases in size, grows into the space between the umbilical vesicle, amnion and sub-zonal membrane, becomes more or less intimately united with these structures, but more especially with the sub-zonal membrane. As the walls of the pleuro-peritoneal cavity grow and close in towards the ventral mesial line a part of the allantois becomes enclosed within the abdominal chamber and forms the urinary bladder, whilst a part remains outside this cavity and forms a dilated bag, the sac of the allantois, which contains the allantoic fluid. This sac keeps up its continuity with the urinary bladder by a slender tubular stalk, the urachus, which passes through the abdominal wall at the umbilicus. The wall of the sac of the allantois, having reached the inner surface of the sub-zonal membrane, spreads itself over the whole, or a great part of that surface, and forms the inner layer of the permanent chorion. As the allantois is a vesicular outgrowth from the alimentary tube, its walls are necessarily derived from the same layers of the blastoderm. It is lined by an epithelium continuous with the epithelial lining of the intestine, which is derived from the hypoblast, whilst external to the epithelium is a layer of connective tissue continuous with and derived from the splanchno-pleure layer of the mesoblast. In this outer connective-tissue layer blood-vessels are developed, which are continuous with

the abdominal vessels of the embryo. When the sac of the
allantois reaches the sub-zonal, or external, layer of the permanent chorion, the blood-vessels are necessarily brought there at
the same time, and as the allantois grows to form the inner
layer of the chorion its blood-vessels become the vessels of the
permanent chorion, or the vessels of the fœtal placenta. These
vessels increase in size and numbers with the growth of the
chorion, their trunks become the umbilical arteries and vein,
they enter into the formation of the umbilical cord and pass
into the abdominal cavity at the umbilicus to join the aorta
and post caval vein.

There are thus four structures, of which one is developed
at the periphery of the ovum, whilst the remaining three

Fig. 1.

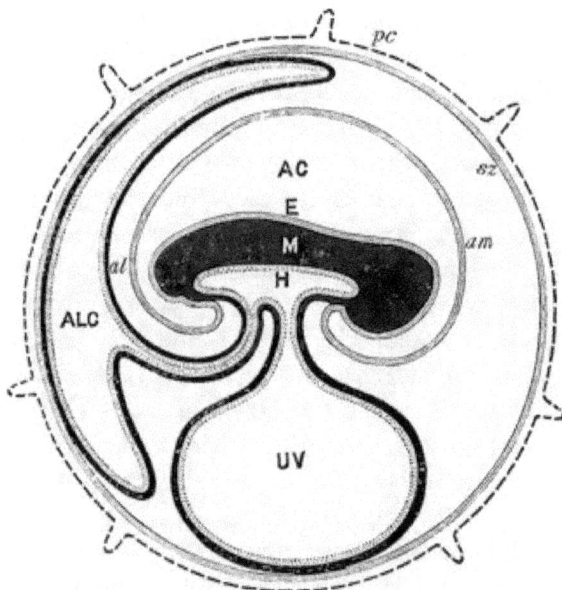

Diagram of the fœtal membranes. Structures which either are, or have been
at an earlier period of development, continuous with each other are represented
by the same character of shading.   pc, primitive chorion with its villi.
sz, sub-zonal membrane. E, epiblast or cuticular covering of embryo continuous
with am, the amnion. M, mesoblast forming the bulk of the body of the
embryo, and prolonged into the wall of the sac of the Allantois al, and into the
wall of the Umbilical Vesicle, UV. H, hypoblast forming the epithelial lining
of the intestinal tube and prolonged into the wall of the sac of the Allantois and
of the Umbilical Vesicle. AC, cavity of Amnion. ALC, cavity of Allantois.

ultimately reach the periphery, which from their position may enter into the formation of the persistent chorion, or outer envelope of the fœtus : viz. the zona pellucida, the sub-zonal membrane, the allantois, and the umbilical vesicle. The structureless zona, with its simple structureless villi, which together form the primitive chorion, very early disappears, either by becoming incorporated with the sub-zonal membrane, so as no longer to be recognised as an independent membrane, or by becoming absorbed. The sub-zonal membrane, or serous envelope of the ovum of Von Baer, originally continuous with the amniotic folds, and through them with the epiblast layer of the blastoderm, persists, and forms the external or epithelial non-vascular layer of the persistent chorion, and of the villi which grow from it. The allantois forms the inner or vascular layer of the persistent chorion, and the vascular matrix of the villi ; but in some of the *Rodentia* the umbilical vesicle reaches and remains in contact with a limited portion of the sub-zonal membrane, to which it conveys connective tissue and blood-vessels. The persistent chorion, therefore, is a compound membrane, produced secondarily during the process of development by the union of the sub-zonal membrane, from which its epithelial layer is derived, with the allantois from which it derives its blood-vessels and connective tissue ; and like the embryo itself it arises from the layers of the blastoderm.

The allantois undergoes considerable modifications in both size and general disposition in the different orders of mammals. In many mammals, as the *Pachydermata, Ruminantia, Cetacea, Carnivora, Pinnepedia*, the Lemurs and the Scaly Anteaters (*Manis*), the sac of the allantois persists as a distinct chamber, and contains a considerable quantity of allantoic fluid. Of the opposite walls of the sac, the one lines, to a greater or less extent, the inner surface of the permanent chorion, which is coated therefore by the epithelial lining of the allantois ; the other invests, to a greater or less extent, the outer surface of the amnion, which is covered therefore, in whole or in part, by the wall of the sac of the allantois. As a consequence of this arrangement the bag of the amnion in these mammals is altogether or in great part separated from the inner surface of the permanent chorion.

In *Orca gladiator*, when the allantois reached the chorion, it became prolonged both to the right and left in the form of a comparatively wide, almost cylindrical, sac-like horn. The left horn-like prolongation extended to seven inches from the end of the left horn of the chorion, where it formed a cul-de-sac. The right horn-like prolongation entered the right cornu, but did not reach to within twenty-one inches of the tip of the right horn of the chorion. A large part of the wall of the allantois was in apposition with the amnion, but, opposite the abdominal aspect of the embryo, the wall of the allantois was united with the inner surface of the permanent chorion by gelatinous connective tissue.

Fig. 2.

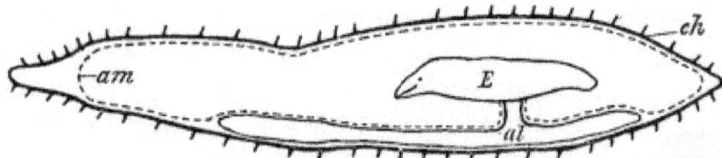

Outline diagram of the arrangement of the fœtal membranes in *Orca gladiator*. *E*, the embryo. *ch*, the permanent chorion with its villi. The dotted line *am* represents the amnion. *al*, the wall of the allantois.

So large and persistent is the sac of the allantois in the ordinary *Ruminantia*, in the *Camelidœ*, *Tragulidœ*, *Solipeda*, and cloven-hoofed *Pachydermata*, that M. H. Milne Edwards has grouped them together as *Megallantoids*: whilst he has placed the *Carnivora* and *Pinnepedia*, from the smaller size of the sac of the allantois, in a *Mesallantoid* legion of mammals. The *Rodentia*, *Insectivora*, *Cheiroptera*, *Quadrumana*, and Man are characterized by the small size of the sac of the allantois, or even by its complete disappearance as an independent chamber: hence Milne Edwards has grouped them together in a *Micrallantoid* legion[1]. The disappearance of the allantois as a distinct sac is correlated with the great expansion of the amnion, and increase in the amount of amniotic fluid. As the bag of the amnion expands it grows outwards towards the chorion, and in the

[1] *Considérations sur les affinités naturelles et la classification méthodique des Mammifères.* Paris, 1868.

Sloths, the Apes, and the Human Female, reaches, and adheres, through the intermediation of a gelatinous connective tissue, to the inner surface of that membrane, so that when the chorion is cut through, the amnion is at once exposed.

The umbilical vesicle forms a sac situated in relation with the abdominal aspect of the embryo, and continuous with that part of the intestine which subsequently forms the ileum, by a hollow stalk or pedicle, the omphalo-mesenteric duct. In some mammals, as in the Pig, Mare, *Cetacea* and *Ruminantia*, it disappears as development advances, so that no trace of it can be seen in the membranes of an advanced embryo. In others, as the Human Subject, it persists, according to Schultze[1], even up to the end of intra-uterine life, as a minute vesicle at the placental end of the umbilical cord. In the *Carnivora* and *Pinnepedia* again it forms a well-defined sac, in relation to the abdominal aspect of the fœtus, situated between the allantois and amnion, and prolonged laterally into two horns. In the Bitch I have seen its vessels persistent, and forming a well-defined plexus in its wall, though they did not reach the chorion. In the frugivorous Bat (*Pteropus medius*) it is persistent. In the Mole and Shrew it is large, and is described by Prof. Owen[2] as supplying the outer envelope of the ovum with vessels. In the Rabbit, and probably also in other *Rodentia*, the umbilical vesicle reaches the chorion, and forms a large sac in contact with a considerable portion of that membrane, to which it conveys blood-vessels; but the discoid placenta of the Rabbit apparently derives its fœtal vessels only from that part of the chorion which receives its vascular supply from the allantois.

---

[1] *Das Nabelbläschen ein constantes Gebilde in der Nachgeburt des ausgetragenen Kindes.* Leipzig, 1861.
[2] *Comparative Anatomy of Vertebrates,* Vol. III. p. 729.

FORM AND STRUCTURE OF THE CHORION.

THE chorion is the most external of the fœtal membranes. When first formed, it is globular or ovoid in shape. In uniparous mammals, as the Sloths, the Apes, and the Human Female, where the uterus is single, it preserves, with but little alteration, this form throughout intra-uterine life. In uniparous mammals, as the Mare, the *Ruminantia*, the *Cetacea*, and the Hare, where the uterus possesses two horns, the chorion becomes greatly elongated, and extends as far as the ends of both horns. In the Seals, on the other hand, as was pointed out by Barkow in *Phoca vitulina*[1], and as I have seen in *Halichœrus gryphus*, where the uterus is also two-horned, the chorion of the single fœtus is limited to that horn in which the fœtus is situated, and the non-gravid horn, though slightly increased in size, contains only a little mucus. In multiparous mammals, again, as the Pig, the Rabbit, Guinea-pig, and the true *Carnivora*, where the uterus is two-horned, and each horn contains one or more fœtuses, the chorion is elongated, and the ends are rounded, so that it becomes, as is so well seen in the *Carnivora*, lemon or citron shaped: but in the Pig a distinct pouch-like sac of the allantois projects through and beyond each pole of the chorion, and forms the *diverticulum allantoidis* described by von Baer. The chorion, belonging to each embryo, is confined to that horn in which the embryo is situated, though I have seen in the pig's uterus, the chorion surrounding the fœtus situated lowest down in the left horn, passing across the corpus uteri into the opposite cornu.

In its early stage of formation, the permanent chorion is uniformly covered with delicate, slender villi, so that it possesses a soft velvety appearance. In some animals, as the Mare, the *Cetacea*, and *Manis*, even though the chorion increases during gestation enormously in size, the villi continue to cover almost the whole of its outer surface, though here and there a tendency is exhibited for the villi to disappear and to leave a patch of

---

[1] *Zootomische Bemerkungen*, Breslau, 1851.

bare non-villous chorion. In these and the other mammals in which the villi are diffused over the greater part of the chorion, the chorion is a *chorion frondosum*, and the placenta is termed a Diffused Placenta.

But in the majority of mammals a much greater extent of disappearance of the villi from the surface of the chorion takes place as development advances. The villi become aggregated into definite and limited areas, surrounded by smooth portions of chorion, and with this limitation of area the individual villi assume a more complex form. In many orders of mammals the areas in which the villi are collected are so precise as to be almost characteristic. In the ordinary *Ruminants*, for example, the villi are collected into large tufts, called fœtal cotyledons, which cotyledons are scattered in considerable numbers over the surface of the chorion, and the form of placenta produced by this arrangement is called a Polycotyledonary Placenta.

In some other mammals the villi are collected together in a zone-like or annular band, surrounding the equator of the chorion, which, with a similar shaped band in the uterine mucosa, forms a Zonary Placenta. The chorion is quite smooth at and for a considerable distance away from the poles. This mode of arrangement of the villi is best seen in the *Carnivora*. The breadth of this zone, in proportion to the length of the chorion, is much greater in the early than in the later stage of gestation, so that if this membrane be examined, say in the cat, when the ovum is only eight-tenths inch long, an area not more than one-tenth inch long at each pole is free from villi, whilst at the end of gestation, with a chorion seven inches in length, the villous equatorial zone is only 1½ inch in its transverse diameter.

In some mammals the villi persist at one pole of the chorion, and are distributed for some distance over its surface; but the opposite pole, *i.e.* the one next the os uteri, is smooth and free from villi. This arrangement leads to the formation of what has been named a Dome-like, or Bell-shaped Placenta. In *Tamandua tetradactyla* the villi, as M. A. Milne Edwards has described[1], occupy the larger part of the surface of the chorion,

---

but the pole next the os uteri is smooth.   They are not simple
in form, but consist of vascular, very compact vegetations.   In
the *Lemurs* also the same anatomist has described a bell-shaped
placenta[1].   The anterior and middle parts of the chorion, he
says, are almost entirely covered by dense compact villi, constituting a sort of vascular cushion, the result of the confluence
of a number of irregular cotyledons, whilst the villi have almost
entirely disappeared from the opposite pole of the chorion.   In
the Sloths, as I have pointed out[2], the villi are aggregated
together in disc-shaped lobes, which are arranged on the anterior and middle portions of the chorion, but leave the pole
next the os uteri smooth, though vestiges of atrophied villi may
be seen on this smooth surface.

In the *Rodentia*, *Insectivora*, *Cheiroptera*, Apes, and Human
Female, the villi are collected into a comparatively circumscribed
area, which is not unfrequently circular or sub-circular in outline, and forms with the corresponding surface of the uterus a
mass know as a Discoid Placenta.

When the villi first grow out from the chorion they are
short, simple, unbranched processes.   In the Pig they preserve,
as a rule, their simple form, but in most mammals they increase
in length and breadth, and branch in a more or less complicated
manner, so as to present many modifications in shape in different genera, which modifications will be described in the course
of these Lectures.

In my description of the development of the permanent
chorion, I stated that this membrane is compound in structure,
and consists of an outer epithelial layer and an inner vascular
connective-tissue layer.   The villi being outgrowths of the chorion,
are therefore composed of the same structural elements.   When
the villi first appear on the permanent chorion they seem to consist solely of cells, and are probably bud-like outgrowths of the
epithelial layer.   Reitz[3] considers that at their first appearance
they consist merely of protoplasm, in which nuclei are imbedded,
the differentiation into definite epithelial cells not taking place
until a later period.   There is some difference of opinion, whe-

---

[1] *Annales des Sciences Naturelles*, xv. 1872.
[2] *Trans. Roy. Soc. Edinburgh*, 1873.
[3] Stricker's *Handbuch der Gewebelehre*. Article Placenta.

ther these outgrowths are at first solid or hollow. Whichever be the case, there can be no doubt that the vascular delicate connective tissue of the inner layer of the permanent chorion is prolonged into the axis of each villus, so that they ultimately possess a vascular core, with a cellular epithelial investment. It is in the young chorion that the epithelium is most satisfactorily seen. I have examined it both in the Pig and Cat, in both of which it consists of a tessellated layer of squamous epithelial cells, the nuclei in which are distinct. As the animal approaches the full period of gestation, the demonstration of an epithelial layer on the surface of the villi becomes more difficult, and in the chorion at the time of birth it is often impossible to distinguish any tissue in the villus, other than the vascular connective tissue, in which many fusiform or stellate corpuscles are imbedded.

The villi soon begin to assume a more complex form by putting out buds both from their sides and free ends, and if this process is frequently repeated they not only increase in size but become arborescent. The part which first appeared becomes the stem, whilst the primary and secondary buds form the branches with their terminal off-shoots.

Each villus has, as a rule, at least one artery and one vein extending along its axis, which are connected together by a capillary plexus, that is distributed near the periphery of the stem and branches and within the villous buds. In most mammals the buds are large and allow room for many capillaries within them; but the filiform villi of the Mare and Cow are so slender that a single, or at the most a double capillary loop, alone is present, and a similarly simple arrangement exists in the Human villi. The vessels of the villi are branches of the umbilical artery and vein, which form the vascular trunks of the umbilical cord. In the *Carnivora* and *Pinnepedia* the umbilical vessels are not limited in their distribution to the zonary placenta, but slender branches extend as far as the poles of the chorion. In *Cholopus Hoffmanni* the non-placental pole of the chorion contains slender ramifications of the umbilical vessels.

The umbilical cord is the bond of union between the fœtus and the chorion. Its length varies much in different mammals.

It is apparently the longest in the Human subject, where it
has an average length of about two feet, and is also of considerable length in the Apes, the Sloths, in *Orycteropus*, and in
the *Cetacea*, whilst in the *Carnivora* and *Pinnepedia* it is
remarkably short.

The umbilical cord contains two umbilical arteries. Usually
only a single umbilical vein is present, but sometimes, as in
the *Cetacea*, *Ruminantia*, and the Pig, two veins occur; and the
arteries wind in a spiral manner around the vein or veins. In
all mammals in the earlier periods of intra-uterine life the
urachus, which connects the extra-abdominal part of the allantois
with the intra-abdominal part, or urinary bladder, forms an
important constituent of the cord. In those mammals in which
the extra-abdominal part of the allantois persists as a capacious
sac, the urachus remains pervious in the cord, sometimes, as
in the *Cetacea*, being funnel-shaped; at others, as in the
Pinnepedia, forming a very fine tube. Where the allantois
however disappears as a distinct sac, as in the Human subject,
then the urachus lying in the cord atrophies, and cannot be
recognised in the later stages of gestation. The umbilical
vesicle is also a constituent of the cord in the early stages of
formation of the membranes, but all trace of it disappears from
the cord as development advances, except in such mammals
as the *Rodentia*, in which the vesicle persists as a well-defined
sac. The vessels of the cord, the urachus and umbilical vesicle
when present in the cord, are surrounded by a mucous or
gelatinous connective tissue, well known as the gelatine of
Wharton. Virchow and others have described a net-work of
anastomosing, stellate connective-tissue corpuscles in this
gelatinous tissue. Dr Koester, who has carefully examined
its structure[1], states that it consists of a mucous basis substance in which fibrillar threads are interwoven and connected
together. The appearance of an anastomosing network with
thickened nodes is due to a canalicular system in which cells
rich in protoplasm are situated, so that the cord is permeated
by an anastomosing system of juice-canals capable of being
injected, which are identical with the connective-tissue cor-

[1] *Ueber die feinere Structur der menschlichen Nabelschnur.* Inaugural Dis-
sertation. Würzburg, 1868.

puscles of Virchow. Virchow has also pointed out[1] that blood-capillaries, connected with the vessels of the abdominal wall, extend for from 4 to 5 lines into the abdominal end of the cord, but no further. Mr Lawson Tait describes in the Human cord[2] a saccular blood-vascular sinus situated at the omphalic ring, which extends for about 45 millimeters into the true substance of the cord and gives off, at short intervals, trunks, which rapidly break up into capillaries, that enter directly into the canalicular tissue. The cord does not appear to have nerve-fibres passing into it. The several tissues of the cord are invested by the amnion, which forms a sheath around it, and is continuous at the abdominal end with the cuticular covering of the embryo. In the Mare and some other mammals, where the sac of the allantois is persistent, only the abdominal or inner end of the cord is invested by the amnion, the outer or placental end being in relation to the allantois. In the Grey Seal (*Halichœrus gryphus*) I observed that the umbilical vessels in the short cord began to branch two inches from the abdomen, and that in the course of these branches to the placenta they passed between the allantois and the umbilical vesicle, and that some even were suspended in the cavity of the vesicle.

In some mammals the amnion investing the umbilical cord has small bodies connected with it, which project towards the amniotic cavity. In the *Cetacea* they have been noticed by Professors Owen, Rolleston, and other observers. I have especially studied their arrangement and structure in *Orca gladiator*, in which cetacean I found not only the amnion covering the cord abundantly studded with yellowish-brown or olive-tinted corpuscles smaller even than mustard-seeds, but similar bodies were thickly congregated on the amnion, where it covered the funnel-shaped sac and horns of the allantois; whilst on that part of the amnion which was in contact with the chorion, they were much more sparingly distributed. Some of these corpuscles were pedunculated, others were sessile. They had obviously been developed in relation to the attached, and not to the free surface of the amnion, for each was invested by a

---

[1] *Cellular-Pathologie*, 1858, p. 87.
[2] *Proc. Royal Soc. London*, June 17, 1875.

prolongation of that membrane, and, where the corpuscles were pedunculated, the pedicle, which was sometimes one-eighth of an inch long, was formed by a slender filamentous process of the amnion. When sections through these corpuscles were examined with a magnifying power of 480 diameters they were seen to consist of cells, closely packed together: some of the cells were oval, others elongated, others somewhat polygonal in shape. I also found between the amnion and allantois, close to the trunks of the umbilical vessels passing to the left horn of the chorion, three peculiar-looking bodies. The largest, tri-radiate in form, was three-quarters of an inch long by half an inch in its greatest breadth; the others were much smaller, and ovoid in shape. They were in linear series, the largest and highest was one inch and a quarter from the second, and that again half an inch from the third. When the allantois and amnion were separated from each other, they remained attached to the amnion. Each body was invested by a fibrous capsule, and contained a brown pultaceous mass, which when examined microscopically was seen to consist of cells, the great majority of which were circular in form, and had a yellowish colour. The smallest of these cells were of the size of lymph-corpuscles, but as a rule they were twice as large. Most of the cells contained a single nucleus, but some had two or even more, and in a few instances large brood-cells were seen packed with nuclei or young cells. Patches of hexagonal cells were also occasionally seen. These bodies, like the amniotic corpuscles, already described, were developed in relation to the attached surface of the amnion.

In the *Ruminantia* the inner surface of the amnion, more especially where it invests the umbilical cord, is studded with white, or yellowish white, bodies. As seen in the Cow they are sometimes elongated in the form of papillæ, at other times in flattened patches slightly raised above the plane of the amnion. Their longest diameter may reach a quarter of an inch, but usually they are of smaller size. Sometimes a patch or papilla lies singly, but more commonly they are collected into large clusters. They could be readily peeled off the inner surface of the amnion, and in injected specimens, a net-work of fine vessels could be seen ramifying in relation to the deep surface of the membrane, but not entering the papillæ or patches. Hence it seems that these bodies are formed on the free surface of the membrane and are epithelial in position. This view of their nature is borne out on microscopic examination, for they consist of large, flat, nucleated cells, not unlike a squamous epithelium. Many of the cells contained minute spherical strongly refracting particles, about half the size of the nucleus, with well-defined outlines, which formed a prominent appearance in the preparations when many cells were looked at together.

Prof. Claude Bernard has carefully examined not only the arrangement and structure, but the chemical composition of these bodies in the *Ruminantia*[1], and has come to the conclusion that they are the seats of formation of a glycogenic material. He states that they increase in size up to a certain period of intra-uterine life and then degenerate; their organization and development being in inverse relation to the development of the liver; the place of which, as regards its glycogenic function, they seem to supply during the earlier period of intra-uterine life.

In the Elephant Prof. Owen has shewn[2] that the inner surface of the amnion is roughened by brownish hemispherical granules, from 1 line to $\frac{1}{10}$th of a line, commonly about half a line, in diameter, but nothing is said of their microscopic structure. Prof. Rolleston described in the Tenrec (*Centetes ecaudatus*) numerous corpuscles studding the inner surface of the amnion[3]. In form, general arrangement, and microscopic structure, they resembled, he says, *les plaques de l'amnios chez les Ruminants* as described by Prof. Claude Bernard.

It is well known that in the Mare olive-green or brownish bodies, varying in size from a pea to a walnut, or even larger, and called by the veterinarian, "hippomanes", are found lying free in the fluid filling the sac of the allantois. Similar bodies are sometimes seen attached by slender threads to the inner surface of the sac. In a specimen which I examined I found three of these bodies moored to the inner wall of the

[1] Brown-Séquard's *Journal de la Physiologie*, Vol. II. p. 31, 1859.
[2] *Philosophical Transactions*, 1857.
[3] *Trans. Zoological Soc.* Vol. v.

sac by several very slender threads, the longest of which
was four inches, but no thread was more than about half as
thick as the chordæ tendineæ in the human heart. These
threads were continuous with the lining membrane of the
sac, but the nature of their attachment to the body itself
could not accurately be ascertained, owing to the tangled
knot which they formed where they joined it. In another
specimen I found several brownish bodies, about the size of
kidney-beans, situated in the gelatinous tissue which connects
the chorion to the wall of the sac of the allantois. Both the
chorion and allantois were slightly elevated over them, and one
had projected so much towards the sac of the allantois, as to
have caused a thinning and absorption of that membrane; so
much so indeed that a portion of the brownish body was projecting into the sac of the allantois. I then removed the
chorion from the outer surface of this body, which was invested by
a fibrous capsule and contained a light brown pultaceous substance. Examined microscopically this substance was seen to
consist of multitudes of free granules and of cells. The cells
varied in form and size: some were circular and about the
size of lymph-corpuscles, others were two or three times
larger and polygonal. In many of these cells the nucleus was
distinct, but in others no nucleus could be seen. I regard
these bodies as the early stage of formation of the "hippomanes". Arising originally between the chorion and allantois,
as they increase in size they grow towards the latter, cause
absorption of the membrane, and project into the cavity; for a
time they remain attached to the wall of the sac, but the
pedicles of attachment become so attenuated that they at last
give away and the "hippomanes" then lie free in the allantoic
fluid.

In the Pig numerous white spherical bodies of the size
of very small shot are situated in the gelatinous tissue which
connects the chorion to the wall of the sac of the allantois.
They are not to be confounded, as has apparently been done
by some writers, with the white circular spots on the chorion
itself, to be subsequently described. Each spherical body
possessed a distinct capsule, which when squeezed gave way
and allowed a drop of fluid to escape. When examined micro-

scopically, multitudes of circular cells, about the size of lymph-corpuscles, and quantities of granular particles, were seen in this fluid. I agree with Prof. Rolleston in regarding them as in all probability homologous with the hippomancs in the membranes of the Soliped.

In the Elephant also, as Prof. Owen has described, numbers of flattened, oval or subcircular bodies, varying in diameter from an inch or more to half a line, are situated between the chorion and adherent part of the allantois, and project inwards towards the allantoic cavity. Through the courtesy of Prof. Flower I have been permitted to take for microscopic examination a portion of the allantois of this specimen, which, with the other membranes, has been preserved in spirit for many years in the Museum of the College. I found the bodies described by Prof. Owen somewhat shrivelled by the prolonged action of the spirit. They were tough and condensed, and difficult to dishevel with the dissecting needles. Under the microscope they were seen to be composed for the most part of fine fibres, having the aspect of white fibrous tissue, the fibres being cemented together by a homogeneous inter-fibrous material, in which nuclear-looking particles were imbedded, but I did not see the abundant cell-forms and granular matter, such as I have described in the bodies occupying a similar position in the pig and mare.

STRUCTURE OF THE MUCOUS MEMBRANE OF THE UNIMPREGNATED UTERUS.

It will now be necessary to examine the structure of the non-gravid uterine mucosa, so that we may see what the tissues are in which the changes occur which lead to the production of the maternal part of the placenta.

In the Human uterus, and in that of some other mammals, the mucous membrane is smooth : but in many mammals it is elevated into folds. In the Pig these folds are transverse, but usually they are arranged longitudinally, *i.e.* parallel to the long axis of the body of the uterus, or of the uterine cornu. In the Mare, most Ruminants, the Seals, the Rabbit and the *Nycticebus tardigradus*, the longitudinal folds are strongly marked. In the Chimpanzee the folds are more feeble, and separated by shallow intermediate fissures. The thickness of the mucous coat, both absolutely and relatively to the thickness of the muscular coat, varies considerably in different mammals. In the Human uterus, as is well known, the mucosa is relatively very thin, but in *Tragulus javanicus* it is equal in thickness to the muscular coat. In the Pig, Mare, various Ruminants, in *Dasypus kappleri*, and indeed in all mammals where the mucosa is folded, it forms a well-defined layer.

The mucosa is covered on its free surface by an epithelium, which rests on a delicate sub-epithelial connective tissue, in which the utricular glands, the blood-vessels, lymph-vessels and nerves of the mucous membrane are situated. In all placental mammals, so far as yet examined, the cells of the epithelium covering the mucosa lining the cavity of the body of the uterus are columnar in shape, and the free ends of these cells give origin to vibratile cilia which project into the uterine cavity. In the Human subject, and probably in other mammals which possess a cervix uteri, the cells covering the part of the mucosa nearest the body are ciliated and columnar, whilst those which lie nearer the os uteri are, like the epithelium of the vagina, squamous and stratified.

To Malpighi is ascribed the merit of having first recognised the orifices of glands on the surface of the uterine mucosa. He saw them in the cow and sheep, and since his time they have been investigated by several anatomists, amongst whom I may more especially refer to Weber, Eschricht, Sharpey, Bischoff, Leydig, Ercolani and Lott, not only in these but in several other mammals. I have myself examined the utricular glands in the following placental mammals; the Pig, Mare, *Orca gladiator*, Sheep, Cow, *Lama paca*, *Tragulus javanicus*[1], Cat, Bitch, Fox, Badger, *Paradoxurus pallasii*, *Halichærus gryphus*, *Cystophora cristata*, *Manis*, *Dasypus kappleri*, *Cholopus Hoffmanni*, Hedgehog, *Nycticebus tardigradus*, *Macacus rufescens*, *Semnopithecus entellus*, *Ateles gricescens*, *Hylobates agilis*, *Troglodytes niger* and the Human Female.

I shall not in this section enter into a detailed description of the structure and arrangement of the glands in these mammals, but shall limit myself to a statement of a few general observations, as in the course of the Lectures, when the structure of the different forms of placentæ is described, it will be more convenient to speak of them in detail.

The utricular glands are branching tubes, which vary in their length, direction, the amount of branches, and in their relative numbers in different mammals. In many mammals they extend perpendicularly to the plane of the surface for some depth, and then incline with more or less obliquity, or possess a tortuous course. Vertical sections through the mucosa, as in *Dasypus*, *Semnopithecus*, the Sheep, Alpaca, Cat, Badger, Bitch, Hedgehog, Mare, &c. give most satisfactory views of the opening of these glands on the surface, whilst in the deeper layer of the mucosa the glands are repeatedly cut across obliquely or transversely. In *Tragulus*, *Cholopus*, *Nycticebus*, *Macacus*, *Hylobates*, the Chimpanzee and the Human Female, the glands throughout their entire length have a more oblique course, so that it is more difficult to see in vertical sections their communication with the surface. Sometimes,

---

[1] I am indebted to A. H. Garrod, Esq., Prosector to the Zoological Society, for the uterus of this and of most of the rarer mammals named in this list. The uteri were kindly placed by him in spirit of wine immediately after remova. from the body, so that sections of the mucosa could be readily made and submitted to microscopic examination.

as in the Hedgehog and *Dasypus*, the glands divide into two branches near the surface, and then give off no more branches, but in other mammals, as the Pig, Mare, and *Orca*, the glands divide repeatedly in the deeper part of the mucosa, so that numerous branching tubes are connected with a single stem or gland-duct. In the Cat, Bitch, Badger, Seal, Hedgehog, *Semnopithecus*, the glands are so numerous and closely packed together, that the amount of interglandular connective tissue between any two adjacent glands is about equal to the transverse diameter of a gland. In *Cholopus*, *Macacus*, *Hylobates*, Chimpanzee, the glands are more sparingly distributed, so that considerably wider intervals exist between them. The glandular epithelium consists of columnar cells, which project vertically into the gland-tube, but leave a central lumen. In 1852 Nylander and Leydig observed[1] that the free ends of the columnar cells lining the utricular glands in the Pig were ciliated, and in 1870 Lott published a memoir in which he stated that he had observed ciliary movement in the epithelium of the uterine glands of the Pig, Cow, Sheep, Rabbit, Mouse and Bat[2]. Further, he believed it to be probable, from the agreement in characters of the epithelial cells of the Cat, Bitch, Guinea-pig, Mare and Human Female with those of the mammals in which ciliary movements were actually seen, that the cells also in them were ciliated.

The utricular glands are not formed during intra-uterine life, and it is probable that they are not fully developed until the uterus reaches the stage of sexual maturity. I examined the uterus of a fœtal calf, six inches long, and though I found its inner surface thrown into longitudinal folds which were in many places still further subdivided by fissures, and covered by a well-marked layer of columnar epithelium, no trace of utricular glands was to be seen. Again I examined the uterus of a new-born lamb without finding any glands, though the folds of the mucous surface possessed a definite columnar epithelial investment. Both in the calf and lamb the sub-epithelial connective tissue was crowded with corpuscles, many of which had the fusiform shape of the corpuscles of connective tissue, whilst

[1] Müller's *Archiv*, 1852.
[2] Stricker's *Handbuch der Gewebelehre*; Article *Uterus*.

others were spherical in form, and resembled the white corpuscles of the blood.

From the observations of Dr G. J. Engelmann[1], it is clear that in the human uterus also no glands exist at the time of birth. In the third or fourth year the mucosa, which had previously been thin, increases in thickness, and small fossæ lined by an epithelium appear in it, which are apparently the rudimentary glands. After the tenth year the glands assume a tubular form, are more numerous, and extend through the thickness of the mucosa into the muscular coat. In girls of twelve or thirteen well-defined glands are present.

The blood-vessels of the uterine mucous membrane are continuous with those of the muscular coat. In those uteri in which a definite sub-mucous coat exists the vessels ramify in it before entering the mucous membrane. When in the mucosa the larger vessels run almost vertically towards the surface and form a capillary plexus, which invests the utricular glands, and near the surface of the mucosa forms a ring around the mouth of each gland.

The lymph-vessels of the unimpregnated mucosa have recently been investigated by Dr Leopold[2]. He considers that lymph-spaces exist between the bundles of connective tissue, the walls of which are formed by the endothelial cells investing those bundles, and that similar endothelial sheaths invest the glands and blood-vessels. He is of opinion that the glands and blood-vessels may be regarded as traversing the lymph-spaces, and separated only from their cavities by their sheaths invested by the endothelial cells.

Nothing definite is known of the mode of arrangement and termination of the nerves of the uterine mucosa. In the mucous membrane of the placental area of the gravid Grey Seal (*Halichœrus gryphus*), I have seen slender nerve-trunks giving off fine branches, which ramified in the connective tissue, until the finest branches consisted of but one or two fibres.

When the fertilized ovum is received into the cavity of the uterus the mucosa undergoes important changes. It swells up,

---

[1] *American Journal of Obstetrics*, May, 1875, p. 32.
[2] *Archiv für Gynækologie*, VI. Abstract in *Journal of Anatomy and Physiology*, VIII. p. 301.

becomes thicker, softer, and more vascular. Its epithelial
covering usually, though not always, loses its columnar form ;
its glands enlarge throughout their entire length : the interglandular tissue increases largely and rapidly in quantity, by a
multiplication not only of the cells of the surface-epithelium,
but by a proliferation of the corpuscles of the sub-epithelial
connective tissue, so that the glands are separated from each
other by a much greater amount of interglandular tissue than
in the non-gravid state; the blood-vessels not only increase in
numbers but in size. At the same time the free surface of the
mucosa is perforated by multitudes of small openings, easily to
be seen with a pocket-lens. Those openings lead into depressions in the swollen mucous membrane, which are usually
regarded as the dilated mouths of the tubular glands, but
which, as I shall show in the course of these lectures, are the
mouths of crypt-like depressions in the interglandular part of
the mucous membrane. These crypts are for the lodgement
of the villi, which project from the outer surface of the persistent chorion. As the area of distribution of the villi on the
chorion varies very considerably in extent in the different forms
of placenta, the distribution of these crypts in the uterine
mucosa necessarily also varies; for the crypts and the villi are
correlated with each other.

The structure of the placenta in those animals in which it
is said to be diffused will first engage our attention.

STRUCTURE OF THE DIFFUSED PLACENTA.

·In the diffused placenta the villi are distributed over almost the entire outer surface of the chorion, and the uterine crypts exist in a corresponding area of the mucous membrane. This form of placenta is found in the Pigs, the *Solipeda*, the *Cetacea*, in *Manis*, the *Camelidæ*, the *Tragulidæ*, the Tapir, the Hippopotamus, the *Lemurida*, and probably the Rhinoceros.

Pig.—I shall commence by describing the structure of the placenta as I have myself observed it in the common Pig, and shall in the first place speak of the fœtal placenta. The surface of the chorion of a pig, where the embryo was 1·3 inch long, was traversed by multitudes of feeble ridges, visible under low powers of the microscope, but no true villi could be seen. A distinct and compact capillary plexus was present both in the ridges and intermediate parts of the chorion. In the ridges the plexus was elongated, but it formed a polygonal network in the intermediate areas. The polar ends of the chorion were smooth and free from ridges for about three inches from each pole. A uniform layer of squamous epithelial cells, the nuclei in which were distinct, covered the face of the chorion.

In an older specimen, where the fœtus was six inches long, the chorion was thrown into well-marked transverse folds in conformity with the foldings of the uterine mucosa, and was traversed by feeble ridges which were more elevated than in the younger specimen, but still requiring a microscope for their examination. The summit of each ridge was broken up into numerous short, simple villi, just as a mountain-ridge may be broken up into short peaks and summits. In injected preparations these ridges and villi were seen to be very vascular[1]. Scattered irregularly over the surface of the chorion were quantities of circular, or almost circular, slightly elevated spots, which varied in number in a given area. Sometimes as many as 30 were seen in a square inch, in other places not more than

---

[1] For valuable aid in injecting this and the other injected placentæ described in these Lectures, I have to express my thanks to my Museum Assistant, Mr A. B. Stirling.

20. The spots varied in diameter from $\frac{1}{20}$th to $\frac{1}{10}$th inch, occasionally one $\frac{1}{8}$th inch in diameter was seen. In most of the red injected specimens these spots were white and free from colour, as if non-vascular, but when the injection was pushed further they became red also, though not so completely as the rest of the chorion. Examined microscopically each spot was observed to have a minute central depression surrounded by villi, which were the terminal villi of a group of ridges, and it was now seen that the villous ridges were arranged on the chorion with especial reference to these spots; for each spot was a centre from which the ridges radiated outwards as the spokes do from the centre of a wheel, so that it had a star-like arrangement. After proceeding some distance the ridges not unfrequently branched, and adjacent branches joined together to form a network. Hence the villous surface of the chorion may be regarded as mapped out into a number of areas, the centre of each area being a circular spot, from which the villous ridges radiate.

In the more fully injected parts of the chorion a vascular ring was seen to surround each star-like spot. This ring con-

Fig. 3.

Portion of injected Chorion of Pig, as seen under a low magnifying power, to show the minute star-like spot *b*, enclosed by a vascular ring, from which villous ridges, *r, r, r*, radiate. Magnified about 20 diameters.

sisted of a double row of capillaries, which passed from the capillaries of one villous ridge to those in the next adjacent ridges, so that, the vessels in the whole series of ridges radiating from any given spot being connected together, a vascular ring was formed. But it was particularly observed that this ring was not formed of vessels passing between the central ends of the ridges, but a little to their outer side, so that the ends were enclosed by the vascular ring.

The existence of small circular spots whiter than the surrounding parts of the chorion was recognised so far back as 1781 by John Hunter[1]. Von Baer gave a careful description[2] of their star-like arrangement, and described and figured their vascularity; more particularly did he describe the relations of their capillaries to the branches of the umbilical vein, so that if a blue injection be introduced into the umbilical vein, and a red into the artery, these spots would appear as blue stars on a red ground. Eschricht also examined these spots[3] and arrived at a similar conclusion as to their relation to the umbilical vein. M. Flourens described[4] these spots as small discs, which are, he states, a variety of the multiple form of placenta. He would obviously associate them therefore with the fœtal cotyledons of a ruminant. But this mode of regarding them is quite erroneous, for the centre of each spot is non-villous, and when the chorion is *in situ*, as I shall immediately point out, is in contact with a feebly vascular part of the uterine mucosa and not with the highly vascular crypts.

Each end of the chorion, for nearly three inches from the pole, had a smooth non-villous surface, and though possessing considerable vascularity was not so vascular as the villous part of the chorion. Hence the chorion of the pig is not uniformly villous, but the villi, as was indeed known to von Baer, are distributed over the central and not the polar regions of the

[1] *Essays and Observations*, edited by Prof. Owen, Vol. I. p. 198. London, 1861.
[2] *Untersuchungen über die Gefässverbindung zwischen Mutter und Frucht*, p. 9, Fig. 1, 2. Leipzig, 1828.
[3] *De organis quæ respirationi et nutritioni fœtus mammalium inserviunt*, p. 36. *Hafniæ*, 1837.
[4] *Cours sur la Génération*, recueilli et publié par M. Deschamps, p. 147. Paris, 1836.

chorion[1]. Though the structure of the placenta in the pig is simple, like that of the mare and other animals with a diffused placenta, yet, as regards the distribution of the villi, they are arranged in a well advanced ovum as a very broad zonular band which does not reach to within several inches from the poles of the chorion.

On examining the maternal placenta in the gravid uterus of a pig, where the fœtus weighed only 12 grains, the free surface of the mucosa was seen to present an undulating appearance, owing to numerous shallow furrows and fossæ separated by intervening ridgelets. Opening on the surface of the mucosa, the openings being marked by shallow depressions distinct from the furrows above referred to, were the mouths of the utricular glands, and not unfrequently a plug of epithelium projected through the orifice. Each gland-orifice was surrounded by a smooth portion of the mucosa.

In a more advanced specimen, where the fœtus was 6 inches long, the mucosa was thrown into transverse folds. When examined with a pocket-lens fine ridges and furrows were seen, which were adapted to furrows and ridges on the surface of the chorion. When more highly magnified the furrows were seen to be subdivided into shallow crypts into which the villi of the chorion fitted. Scattered over the surface of the mucosa were numerous smooth almost circular depressed spots, free from ridges and crypts, corresponding in size and numbers to the circular star-like spots on the surface of the chorion; and, as von Baer[2] and Eschricht[3] have already described, the star-like elevations of the chorion are adapted to these smooth spots on the mucosa when the two surfaces are in contact. These smooth depressed spots were very definitely circumscribed by the crypts which immediately surround them. In the minutely-injected uterus the ridges and the walls of the crypts were seen to contain a compact capillary plexus, whilst the smooth spots possessed a feeble vascularity, so that they appeared as distinct white spots, surrounded by highly vascular ridges, on

---

[1] *Untersuchungen* (op. cit.), p. 7, Plate, Fig. 1.
[2] *Untersuchungen* (op. cit.), p. 12, and *Ueber Entwickelungsgeschichte der Thiere*, Zweiter Theil, p. 250, 1837.
[3] *De organis*, &c. (op. cit.). Hafniæ, 1837.

the injected mucous surface. Beneath the superficial crypt-layer of the mucous membrane was a well-defined glandular layer, the glands in which were tubular and branched repeatedly, so that each gland-stem or duct had connected with it numerous

Fig. 4.

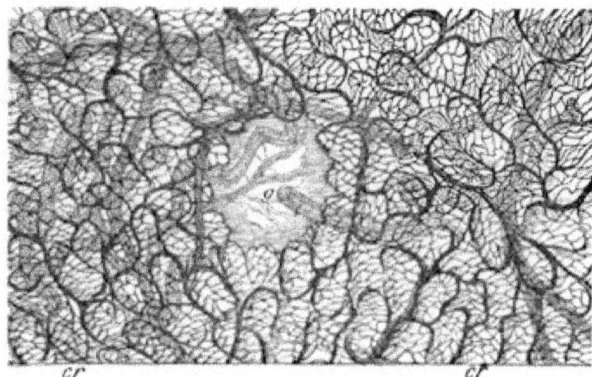

Surface-view of a portion of the injected Uterine Mucosa of a Pig to show a depressed circular spot in which the mouth of a gland g, opens. This spot is surrounded by numerous vascular crypts cr, cr. The branching glands of the glandular layer and the larger vessels lie deeper than the crypts. Magnified same scale as Fig. 3.

branching tubes. The depressed circular spots had a special relation to the ducts of these glands, for opening either in the centre of each spot, or near its border, by an obliquely-directed orifice, was a gland-stem, which could be seen running, some-what tortuously, from the deeper glandular layer of the mucosa to the spot.

Eschricht had in 1837 described the mouths of the utri-cular glands in the pig as situated in small circular spots (areolæ), distinct from the surrounding crypts (cellulæ), which circular spots in injected specimens, owing to their feeble vas-cularity, appeared white, when compared with the highly vascular crypts[1]. These observations by Eschricht seem to have

[1] It should be said that von Baer (Untersuchungen, p. 12) in his description of the depressed spots on the uterine mucosa of the gravid pig, stated that in the middle of each spot a fine vessel arose by an open mouth, which was never injected when the blood-vessels were filled. The true nature of this "vessel" was not however at that time understood by von Baer, for he believed it to be a lymphatic. Subsequently (Entwickelungsgeschichte, Zweiter Theil, p. 250), and after the publication of Weber's researches on the glands of the uterus, he recognised that this so-called vessel was an utricular gland.

been overlooked by most subsequent anatomists. In 1871[1] I described a similar arrangement in the uterus of a pig which I examined. In 1873 a description with a characteristic figure of a spot in an uninjected specimen was given by Ercolani[2]. It is clear therefore that in the pig the glands do not open into the crypts of the gravid mucosa, but into special depressions on the mucous membrane distinguished by a difference in form and in the degree of vascularity from the surrounding crypt-like surface. The crypts therefore are interglandular in position, and are produced by modifications in the interglandular part of the mucous membrane, and not by a dilatation of the gland-orifices themselves. The free surface of the mucosa was covered by an epithelium which also lined the crypts. The epithelial cells were columnar in form, finely attenuated at their deeper end; and not unfrequently I saw an appearance as if cilia projected from the broad end of the cell; but the animal had been too long dead to enable me to determine their presence by vibratile movements. Owing to the shallowness of the crypts and the very short villi of the chorion in the pig the uterine and chorionic surfaces separated from each other with great readiness. As the tubular glands did not open into the crypts their secretion did not come into immediate contact with the general villous surface of the chorion, but with the star-like spots, enclosed by the vascular rings, from which the villous ridges radiated. The mucous membrane of the gravid uterus of the pig differs from that of the non-gravid animal in the following characters: in the presence of a layer of crypts, in the increased size and greater obliquity of the glands, and in the much greater vascularity of the membrane generally.

Mare.—I have not had the opportunity of examining the foetal placenta of the Mare in the early period of gestation, but two specimens in advanced stages have come under my observation. In both, the surface of the chorion presented a soft, velvety, vascular appearance, due to its being almost uniformly covered with vascular villi, even up to the poles. But at each pole, where the chorion was in relation to the orifice of the Fallopian tube, a spot, a fraction of an inch in diameter, smooth and bare

---

[1] *Trans. Roy. Soc. Edinb.* Vol. XXVI. p. 490.
[2] *Mem. dell' Accad. delle Scienze di Bologna.* Plate 2, Fig. 1.

of villi, was present. Opposite the os uteri internum a well-defined bare patch, which in one specimen was about an inch in diameter, having in its centre a faint papillary elevation, was present. Radiating outwards from this patch were five long branching arms, also free from villi, and immediately beyond these some irregularly-shaped bare patches were seen. The surface of the uterine mucosa was also very vascular, except opposite these bare patches on the chorion, and the folds and depressions radiating from the os uteri corresponding to the radiated patch on the chorion had a comparatively slight vascularity. In the chorion from another mare the bare spot opposite the os uteri measured $2\frac{1}{2}$ inches long by from $\frac{1}{2}$ to $\frac{3}{4}$ inch broad, and had distinct bare radii passing off from it. In each specimen an irregular-shaped non-villous patch, about an inch long, was found on a portion of the chorion not in relation to a uterine orifice. Though in the mare the villi seem to the naked eye to be closely set over the surface of the chorion, yet when examined with low powers of the microscope they are seen to be arranged in brush-like clusters or tufts, separated by narrow non-villous intervals, an arrangement the importance of which will appear when the corresponding surface of the mucosa is described. The tufts are like minute fœtal cotyledons, and the villi in each tuft are filamentous in shape and contain a loop of capillary blood-vessels. That part of the chorion which occupied the corpus uteri was thrown into occasional longitudinal folds, but at the horns, more especially near the free ends, the longitudinal folds were numerous and closely adapted to similar folds of the mucous membrane.

I have not yet been able to procure the uterus of the mare in an early period of gestation. But on inspecting with a simple lens the surface of the uterine mucosa of a mare which had reached an advanced stage of pregnancy[1], I found it subdivided into multitudes of irregular polygonal areas, varying in diameter from $\frac{1}{12}$th to $\frac{1}{20}$th inch, by slender ridges, which anastomosed with each other so as to have a reticulated appearance. In injected preparations the ridges were seen to be less vascular than the areas which they enclosed,

[1] I am indebted to J. R. W. Dewar, Esq., V. S. of Midmar, Aberdeen, for this specimen.

and consequently they were more readily recognised in injected
than in non-injected portions of the mucosa. But in addition
the ridges were smooth on the surface, whilst the enclosed
areas possessed a delicate punctated appearance. When more
highly magnified each area was seen to be subdivided into
multitudes of crypts, which passed more deeply into the mucosa
than in the pig. The arteries and veins of the mucosa occupied
the ridges, and broke up into small branches which ended in a

Fig. 5.

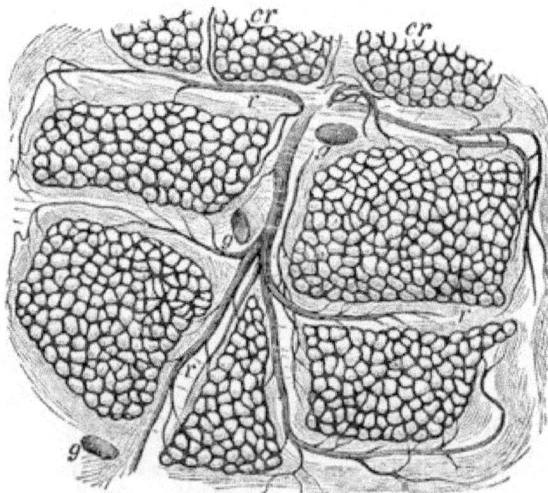

Surface view of the injected uterine mucosa of a gravid Mare, to show the polygonal crypt-areas cr cr, with the anastomosing capillary network surrounding the orifices of the crypts. g, g, g, mouths of three utricular glands
opening on the surface of the ridges r, r, r, which separate adjacent crypt-
areas from each other. Magnified about 20 diameters.

compact capillary plexus situated in the walls of the crypts,
the artery in each ridge giving off branches to the crypt-areas
between which that ridge was situated. The capillaries also
surrounded the orifices of the numerous crypts opening on the
surface of each area, and formed an anastomosing network, the
meshes of which were not circular, but rather polygonal in
form.

The glandular layer of the mucosa contained numerous
branched tubular glands. Each gland possessed a stem or
duct which ascended almost vertically between the crypt-

areas, and opened on the summit of a ridge by a circular or oval aperture, which was usually situated at a spot where convergent ridges became continuous with each other. In the mare, therefore, as in the pig, the utricular glands do

Fig. 6.

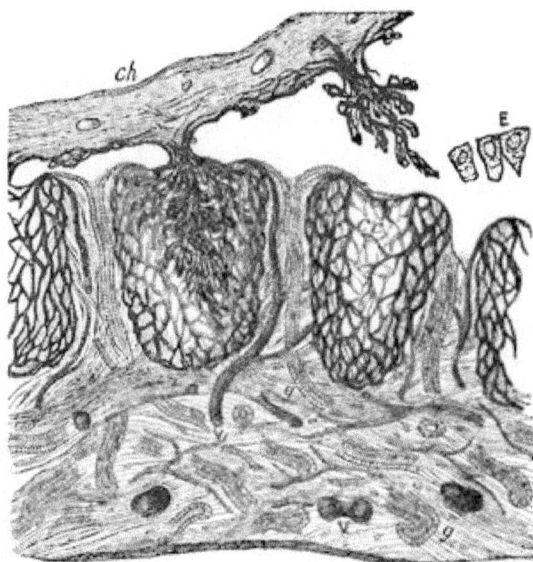

Vertical section through the injected placenta of the Mare. *Ch.* the chorion with its villi, partly *in situ*, and partly drawn out of the crypts, *cr.* *E*, loose epithelial cells which formed the lining cells of a crypt. *g, g*, the utricular glands. *V, V*, the blood-vessels of the mucosa imbedded in the connective tissue.

not open into the crypts; but on definite surfaces of mucous membrane between the crypts, so that the crypts are interglandular in position, and produced by changes in the interglandular part of the mucosa. The demonstration of the want of any communication between the utricular glands and the crypts in the mucosa of the gravid mare, and the consequent interglandular position of the crypts, was made a few years ago by Prof. Ercolani[1] of Bologna.

The surface of the mucosa and of the wall of the crypts was covered by an epithelium, which when examined *in situ* showed

---

[1] See his *Mémoire sur les Glandes Utriculaires de l'Utérus.* French translation. Algiers, 1869, and the figures in Plates 3 and 5.

a polygonal pattern, like the broad free ends of columnar epithelium-cells. When the cells were teased asunder, some were seen to have the elongated form of ordinary columnar epithelium; others were so swollen out that their length but little exceeded their breadth; whilst others were irregular in shape. The protoplasm was distinctly granular, more especially in the irregularly-shaped cells, which resembled in appearance the large granulated cells of the serotina as seen in the higher mammals. Numerous cells exhibiting forms transitional between ordinary columnar epithelium and serotina-cells were seen, so that the large granulated cells of the serotina are to be regarded as a modified epithelium.

The filiform villi of the tufts of the chorion occupied the crypts in the mucosa, which represented therefore the maternal cotyledons of a ruminant animal, and the size of the crypt-areas corresponded to the size of the tufts. The ridges between the areas filled up the intervals between the tufts. The secretion of the uterine glands was not poured into the crypts, so as to come into immediate relation with the villi, but opposite the inter-villous surface of the chorion. The villi of the chorion were so closely fitted into the uterine crypts, that, in the specimen of the gravid uterus near the full time, it required a little force to draw the villi out of the crypts. As in the pig, the gravid mucous membrane differed from the non-gravid in the presence of a layer of crypts, in the increased size and greater obliquity of the glands, and in the greater vascularity of the membrane generally.

Cetacea.—The diffused distribution of the villi over the surface of the chorion in the *Cetacea*, and the velvety appearance due to this arrangement, have been recognised by several anatomists, from observations made more especially on the common porpoise. In 1869 I examined several square feet of the chorion of *Balænoptera Sibbaldii*[1], and showed that the whalebone whales agreed with the toothed whales in the diffused distribution of the villi. But in the *Cetacea*, as in the pig and mare, patches of chorion bare of villi are also present. Some years ago Prof. Rolleston pointed out[2] that in a specimen

---

[1] *Trans. Roy. Soc. Edinburgh*, 1870.
[2] *Trans. Zool. Soc.* p. 307, 1866.

(species unknown), which he examined, a bare spot was situated at each pole. In a gravid *Orca gladiator*, which I examined in 1871[1], I found not only the polar bald spots, about the size of kidney-beans, but a stellate non-villous surface, nearly the size of a crown piece, and with several bare lines radiating from it opposite the os uteri; the arrangement corresponding very

Fig. 7.

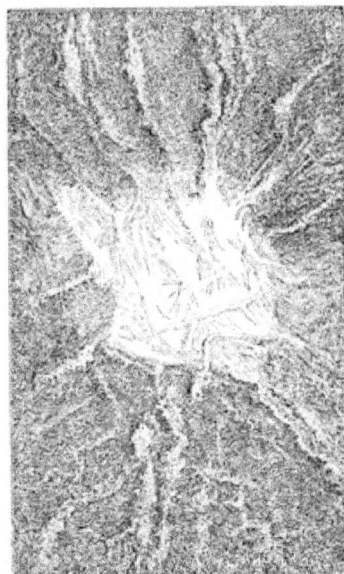

Stellate non-villous portion of the Chorion of *Orca* surrounded by vascular villi, opposite the os uteri. About half the size of nature.

closely with that just described in the chorion of the mare in the same locality. These large radiating bare spots in the mare and cetacean are exaggerated representations of the small radiating spots, so abundantly distributed over the chorion of the pig.

The chorion of *Orca* was thrown into longitudinal folds, more especially near the poles. The villi with which it was covered were seen when examined microscopically to vary in number and in arrangement in different parts. Sometimes they were set in rows and formed parallel ridges: at others

[1] *Trans. Roy. Soc. Edinb.* 1871.

they were collected into tufts, irregular in form and size, which
sometimes consisted of two, three or four villi, but frequently
of a larger number. Solitary villi were also met with in the
irregular intervals between the tufts and ridges; and it was
not uncommon, as Eschricht had observed in *Phocæna*[1], to see
short stunted simple villi projecting from the general plane of
the chorion. The tufts not unfrequently swelled out into a
branching crown, which, to adopt Eschricht's description of
the shape of the villi in *Phocæna*, formed a miniature repre-
sentation of the head of a cauliflower. The secondary villi of
a tuft, as well as the simple villi, were club-shaped.

The basis substance of each villus consisted of a delicate
connective tissue, in which numerous spheroidal and fusiform
nucleated corpuscles were imbedded; some of these corpuscles
were in the walls of the finer blood-vessels, but others were
proper to the tissue itself. A layer of spherical or ovoid corpuscles was situated immediately within the free surface of
each villus, and not unfrequently the periphery of the villus
was slightly elevated immediately above the individual corpuscles, so that the outline of the villus had a gently undulating
appearance. These corpuscles I have named from their position the sub-epithelial corpuscles of the villus. In their form
and appearance they were not unlike the white corpuscles of
the blood, and it is possible that they might have migrated out
of the blood-vessels into the connective tissue of the villus.
The villi contained a distinct capillary network. Not only in
the *Orca*, but in the pig and mare, the capillaries of the
chorion were not limited to the villi, but an extra-villous capillary network, which freely anastomosed with the intra-villous
capillaries, was situated beneath the general plane of the chorion. The blood in its passage from the terminal twigs of the
umbilical artery to the umbilical vein had to flow not only
through the capillaries within the villi, but through the extra-
villous network, from which the rootlets of the vein arose.

Eschricht in 1837[2] and Stannius in 1848[3] described the
presence of numerous little recesses on the free surface of the
uterine mucosa of the gravid porpoise, in which the villi of the

---

[1] *De Organis*, &c., p. 6.   [2] *De Organis*, &c.
[3] Müller's *Archiv*, p. 402, 1848.

chorion were lodged. In 1871[1] I had the opportunity of dissecting the gravid uterus of *Orca gladiator*, and of determining much more minutely than had previously been done, the structure of the gravid uterine mucosa in this order of mammals. The free surface of the mucosa had a delicate reticulated appearance, due to an anastomosing arrangement of slender bands of the membrane. Sometimes a division of the surface into irregular polygonal areas was seen, at others its surface was traversed by an elongated ridge and furrow

Fig. 8.

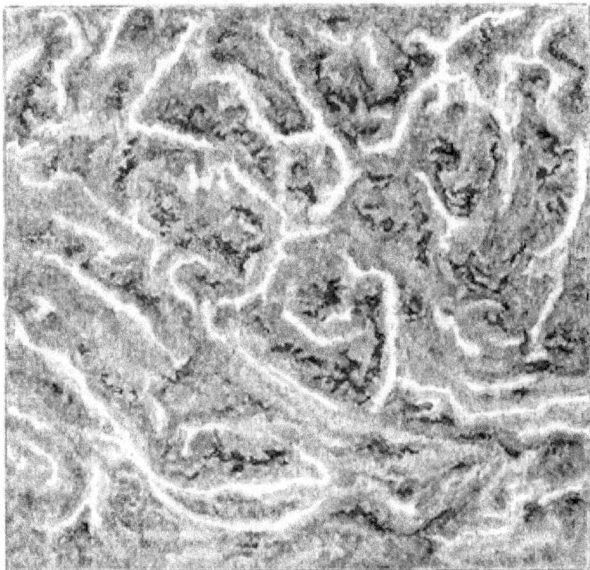

Surface-view, under a low power of the microscope, of a portion of the uninjected uterine mucous membrane of *Orca gladiator*.

arrangement. The polygonal areas and the furrows were subdivided by more delicate bands into small crypt-like compartments, and the intermediate ridges and bands of the mucous membrane were not unfrequently covered by similar crypts. In some places the crypt-like recesses were so closely crowded together that the surface of the mucosa had a honey-comb

[1] *Trans. Roy. Soc. Edinburgh*, p. 467, 1871.

appearance. As a rule the fundus of each crypt was more dilated than its mouth, so as to adapt it to the form of the club-shaped villi.

Beneath the crypt-layer was the glandular layer of the mucous membrane. The glands, as in the pig and mare, were tortuous tubes and branched repeatedly. Sometimes they bifurcated, at others three branches arose close together, which in some instances possessed considerable length, but in others formed short diverticula. Each branch possessed an almost uniform diameter, but the stem of the gland-tube was wider than its various branches. In vertical sections through the membrane only short lengths of any given gland could be traced, for it was frequently divided, sometimes longitudinally, at others, obliquely or transversely.

The mode of termination of the glands on the surface of the mucosa was much more difficult to determine than in the pig and mare, but after repeated examinations, both of vertical sections through the membrane, as well as of surface-views, I came to the conclusion that the tubular glands opened into the bottom of some of the crypts[1]. In many of the vertical sections undoubtedly no gland could be traced into a crypt, a circumstance not to be wondered at when the oblique direction of the stem of the gland-tube is considered; but occasionally a gland-stem could be traced to the bottom of one of the more deeply situated crypts in the mucosa. More satisfactory evidence of the opening of glands into certain of the crypts was obtained by the examination of the free surface of the mucosa. In looking into the crypts through a binocular microscope, I occasionally saw an opening at the bottom of a crypt; the direction of this opening was usually oblique, corresponding indeed with the obliquity of the gland-stem as it reached the surface, and through this opening a little plug, apparently formed either of epithelial cells, or of the coagulated secretion of the gland, was sometimes seen to project into the crypt. Owing to the complexity of the free surface of the mucosa, from

[1] I am not prepared to say that on a surface so extensive as the mucous membrane of the gravid uterus in *Orca* there may not be here and there a spot free from crypts at which a tubular gland may open, but I did not succeed in finding one.

the multitude of crypts, it was not possible to say how many gland-tubes opened in a given area. It was evident however that numerous crypts, which did not have glands communicating with them, lay around those crypts into which the glands opened. It was also clear that the number of gland-stems was very much smaller than that of the tubes in the deeper portion of the gland-layer, so that the number of branches connected with each stem was considerable.

As the crypts were obviously very much more numerous than the ducts of the glands, and as those crypts into which

Fig. 9.

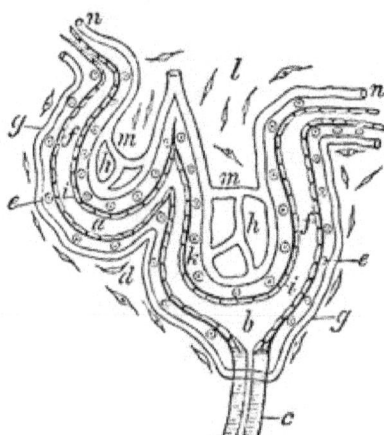

Diagrammatic section through the placenta of *Orca gladiator*. *a.* cup-shaped crypt. *b.* funnel-shaped crypt. *c.* tubular gland-stem with its epithelial lining. *d.* fusiform and *e.* spheroidal sub-epithelial connective-tissue corpuscles. *f.f.* epithelial lining of crypts. *g.g.* maternal capillaries in the walls of the crypts. *h.h.* chorionic villi occupying the crypts. *i.i.* epithelial covering of the villi. *k.* spheroidal and *l.* fusiform corpuscles of the villi. *m.m.* intra-villous fœtal capillaries continuous with *n.n.* extra-villous capillaries. The space represented between the fœtal epithelium *i.i.* and the maternal epithelium *f.f.* is to give distinctness to the diagram, for in the placenta itself the two epithelial surfaces are in close apposition.

tubular glands opened were deeper and more funnel-shaped than those into which glands did not open, I was led to divide the crypts into two groups, non-glandular cup-shaped crypts, and glandular funnel-shaped crypts. The relation of the glands to the funnel-shaped crypts seemed to justify the inference that these pouch-like depressions in the mucosa were (as was stated, by Dr Sharpey, to be the case in the pits or "cells" on the surface of the uterine mucosa in the gravid bitch) the mouths of the glands somewhat enlarged and widened. But however this might be the case with the funnel-shaped crypts it obviously could not be so with the cup-shaped crypts, which were interglandular in position, and, as in the pig and mare, could only have been produced by changes in the interglandular part of the mucous membrane. I guarded myself however against too absolute an acceptance of the view that the funnel-shaped crypts were merely the widened mouths of the glands by stating (op. cit. p. 501) that they, like the cup-shaped crypts, may be formed by a folding of the greatly hypertrophied mucous membrane : only in the one case the hypertrophy and folding take place between the glands, in the other at the mouth of the gland itself. The difference between the mode of opening of the glands in Orca on the one hand, and in the pig and mare on the other, seemed to be this : that in Orca the free surface of the mucosa was much more uniformly crypt-like than in the pig and mare, so that there were no intermediate surfaces destitute of crypts on which the glands could open, whilst in the pig and mare the crypts were collected into definite areas, with distinct smooth surfaces intermediate to them. This more compact arrangement of the crypts in Orca corresponds with the more crowded condition of the villi on the surface of the chorion. That the whole series of crypts however in the cetacean uterus, as in the pig and mare, are to be regarded as interglandular formations, is supported by the observations of Eschricht on the mucosa of the gravid porpoise. For he states (p. 35) that in this animal the glands open on the surface of the mucous membrane, not into the "cells" in which the villi are lodged, but into separate shallow areolæ; that in the porpoise, as in the pig, these areolæ are much less vascular than the surrounding crypts; and that for so great a multitude of gland-ramifications there are not more openings on the surface of the mucous coat than in the pig.

The walls of the crypts and the interglandular connective tissue in Orca contained numerous nucleated corpuscles. In the interglandular tissue they were mostly fusiform, but in the walls of the crypts a distinct layer of globular lymphoid-looking corpuscles was seen close to the free surface, which was not unfrequently elevated, in a sinuous outline, immediately superficial to these corpuscles, which may, from their position, be called the sub-epithelial corpuscles of the crypts. It is not improbable that these corpuscles may have migrated through the walls of the adjacent capillaries. The walls of the crypts were very vascular and contained a compact capillary network. Owing to the closer arrangement of the crypts, the surface of the mucosa generally presented a more uniform vascularity than in either the pig or mare. In all these animals indeed the great vascularity of the crypts was one of the most striking features in the structure. The capillaries in the walls of all the crypts belonging to the same group formed a continuous network, and in the *Orca*, owing to the more uniform crypt-formation, the capillaries of one group freely anastomosed with those of adjacent groups. The capillaries in the crypt-walls in each animal formed a series of anastomosing festoons, and usually a distinct capillary ring surrounded the mouth of each crypt. Moreover there was a great difference between the vascularity of the crypts and that of the deep layer of the mucosa in which the glands were situated, for the vascularity of the latter was not more than may be seen in connection with the tubular glands of the intestine. The crypts in *Orca* were lined by a well-defined layer of epithelium-cells, which closely followed the various irregularities of the mucous surface. The free ends of the cells were polygonal, often hexagons, or pentagons, though sometimes elongated into a pyriform shape. In my original memoir on the placentation of *Orca*, I stated that they had the appearance of a pavement-epithelium, though they were not larger than the broad free ends of the cylindrical epithelium lining the glands. I have since re-examined this layer of cells, and have now come to the conclusion that they cannot be associated with either the pavement-epithelium (*i.e.* if we employ the term as equivalent to squamous), or with the cylindrical epithelium. The cells are neither sufficiently elongated for the one, nor flattened for the other, but have an intermediate or transitional form.

The villi of the chorion fitted into the crypts, but were easily extracted from them. Only those villi which occupied the funnel-shaped crypts were brought into immediate contact with the secretion of the tubular glands, but in all the crypts the villi were in contact with the epithelial lining.

I have also been so fortunate as to acquire the gravid uterus of a Narwhal (*Monodon monoceros*), the fœtus in which was 5 ft. 5 in. long[1]. The uterus, as in *Orca*, consisted of two horns united together in the corpus uteri, and the fœtus lay in the left horn. The chorion extended from the tip of the left to that of the right horn, as in the *Orca* and the Mare. The chorion immediately enveloping the fœtus was not folded: but in the whole length of the right, and at the free end of the left horn it was raised into strong longitudinal folds, which corresponded in reverse order with a similar series of folds of the uterine mucosa radiating from the orifices of the Fallopian tubes. At the os uteri the mucosa was also thrown into strong folds which radiated for some distance into the gravid cornu, and in some parts of their length projected as much as three inches from the general plane of the mucosa, though at the os they had not more than one-half that projection : the chorion in apposition with this part of the mucosa was also folded.

Except in the localities to be immediately specified the whole of the extensive surface of the chorion was so covered with vascular villi that, by the naked eye, at least, no non-villous intervals were recognised. The chorion was so adherent to the uterine mucosa that gentle traction was needed to draw them asunder, and, as the one was peeled off the other, the villi were seen to be drawn out of multitudes of crypts opening on the free surface of the mucosa.

The chorion, in apposition with the os uteri and surrounding folds of mucous membrane, was for the most part non-villous, and presented a smooth, feebly vascular appearance, which contrasted strongly with the adjacent villous vascular chorion. This smooth spot was irregular in form, measured six inches by four, and from it narrow bands of smooth chorion radiated outwards from two to three inches between the villous-

---

[1] I am indebted for this valuable specimen, through the kindness of my friend Mr C. W. Peach, to Mr John Maclauchlan, Chief Librarian and Curator to the Free Library, Dundee. For additional details as to the position of the fœtus, &c., I may refer to *Proceedings Royal Society, Edinburgh*, Feb. 7th, 1876.

covered folds: it was similar to but much larger than the corresponding spot in the Mare, and in *Orca*. Small isolated patches of villi were scattered irregularly over the surface of this smooth spot. A smaller non-villous spot $1\frac{3}{4}$ in. by $\frac{3}{4}$ in. was situated three inches from this large bare surface. The uterine mucosa opposite these smooth portions of the chorion, though folded, was smooth and free from crypts, except where the isolated patches of villi were in apposition with it. Radiating for about one inch from the pole of the chorion in the gravid horn were narrow non-villous bands separated by intermediate villous surfaces. The bands were in apposition with the folds of the mucosa free from crypts, which radiated from the orifice of the Fallopian tube. In the non-gravid horn the chorion was feebly vascular, and devoid of villi, for about five inches from its pole; and even for a greater distance the villi were irregularly scattered, so that well-defined smooth spots could be traced as far as ten or twelve inches from the pole.

When examined microscopically the villi were seen to be arranged in tufts, which varied in size and in the number of villi. Some tufts had not more than two or three villi, but more usually numbers were collected together; though not unfrequently short single villi arose from the chorion in the intervals between the tufts. They had, as a rule, the same shape as those in *Orca*, but some were seen to divide into filiform branches. They contained a capillary network, and a layer of extra-villous capillaries ramified beneath the inter-villous surface of the chorion.

The free surface of the uterine mucosa was, as in *Orca*, pitted with multitudes of recesses and furrows, which again were subdivided into innumerable crypts. In the polar regions of the cornua and in the corpus uteri the mucosa was more spongy and succulent than in the greatly distended part of the gravid cornu, in which the mucosa was obviously more stretched, so that the pits and furrows were almost obliterated, and the crypts opened on the general plane of the mucosa. In their general arrangement and in the vascularity of their walls the crypts in the Narwhal so closely resembled the crypts in *Orca* that no special description is required. But the layer of cells lining the crypts was a well defined cylindrical epithelium, many of the cells of which were however so swollen that the breadth almost equalled the length.

Scattered over the surface of the mucosa in the more distended part of the cornu were numerous smooth, depressed, ovoid or circular spots, the largest of which was not more than $\frac{2}{10}$ths inch in diameter, though as a rule they were less than $\frac{1}{10}$th inch, so as almost to escape observation with the naked eye. 25 or 30 of these spots were on an average in each square inch ; and each was surrounded by a minute fold of the mucosa subdivided into crypts. The corresponding surface of the chorion possessed occasional smooth spots from $\frac{1}{10}$th to $\frac{2}{10}$ths inch in diameter, surrounded by villous tufts, which were in apposition with the depressed spots on the mucosa, but they had not the definite stellate form which one sees in the corresponding spots on the chorion of the pig. The extra-villous layer of capillaries ramified beneath these non-villous spots of the chorion. In the succulent parts of the mucosa the depressed spots could not be seen with the naked eye, and were found only after a careful search with a pocket-lens at the bottom of some of the pits or trenches in the membrane.

The glandular layer of the mucosa corresponded so closely in the direction and mode of branching of the tubular glands with the corresponding layer in *Orca*, that I need not give a special description. The glands were subjacent not only to the crypts, but to the smooth depressed spots, and as these spots bore so close a resemblance to the smooth spots on the mucosa of the pig, I was naturally led to examine if the mouths of the utricular glands opened there. In one instance I saw a tube lined by epithelium lying obliquely beneath the membrane of a spot, and opening near the middle by a distinct orifice bounded by a crescentic fold of the membrane. In a second instance the end of a gland appeared from under cover of the surrounding crypts, and then seemed to open by an obliquely directed mouth near the edge of the spot. But upwards of 30 other spots examined with equal care gave no evidence of gland-mouths opening on them. Hence it would appear that these smooth surfaces on the mucosa are by no means necessarily associated with the mouths of the utricular glands, and one is disposed to conclude that the gland-orifices are usually concealed amongst the crypt-like foldings of the mucosa. The very much greater number of the crypts, than of the gland-stems, negatives the idea of the crypts being merely dilatations of the mouths of the glands, so that in the Narwhal, as in the Pig, Mare and *Orca*, the crypts are to be regarded as inter-glandular in position.

Through the kindness of Dr Allen Thomson I have had the opportunity of examining a portion of the uterus of a Narwhal, in his possession, at a much earlier stage of gestation, the fœtus in which was only $3\frac{1}{4}$ inches long. The free surface of the mucosa was gently undulating and traversed by shallow furrows, but no definite crypts could be seen. The gland-tubes were remarkably numerous, tortuous and branching. I made a comparative measurement of their size, with that of the glands in my much more developed specimen, and found them to have only one-half the transverse diameter. The gland-stems inclined obliquely to the plane surface of the mucosa, on which their orifices could occasionally be seen. The free surface of the chorion was not villous, but traversed by faint ridges, which without doubt fitted into the shallow furrows of the mucosa. Patches of epithelium-cells could be seen covering the surface of the chorion.

It is clear, therefore, that in the Narwhal, as I have previously described in the pig, the villi do not form on the surface of the chorion, nor the crypts on the surface of the mucosa, until the embryo has reached a stage of development in which its body, though small, has assumed a form which enables its ordinal characters to be recognised.

The amnion in my specimen formed an immense bag, which reached to 5 inches from the free end of the left horn of the chorion, but did not pass into the chorion occupying the right uterine cornu. It was so closely adherent to the inner surface of the chorion, that when that membrane was divided the sac of the amnion was opened into. Opposite the abdominal aspect of the fœtus the wall of the allantoic sac was in contact with the chorion, and the amnion was reflected over its surface. The allantois formed a large funnel-shaped sac at the bifur-cation of the funis, and was prolonged along the adjacent surface of the chorion to within two inches of its free end in the left cornu, and nine inches of its free end in the right cornu.

Fig. 10.

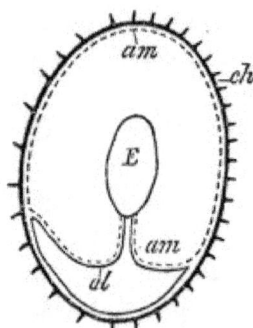

Diagrammatic transvorse section through the fœtus and membranes to show the relations of the allantois to the abdominal aspect of the fœtus, and the relations of the amnion to the chorion. *E.* ombryo. *ch.* the villous chorion. *al.* the allantois, the letters are in the sac. The dottod lino *am. am.* represents the amnion, the letters are in the sac.

The amnion investing the cord was studded with yellowish-brown corpuscles, similar to those that I have described in *Orca gladiator* (p. 23); similar corpuscles were scattered in considerable numbers over the amnion covering the allantois, and a few were seen on the amniotic lining of the chorion. But in addition, numerous dull white corpuscles were found, some of which were slender rods, from $\frac{1}{10}$th to $\frac{4}{10}$th inch long, and arranged end to end like the links of a chain; whilst others were globular, like small shot. The rods were most numerous on the abdominal half of the cord, the globules at and near its bifurcation. These corpuscles were, like those of a yellowish-brown colour, covered by the amnion. When examined microscopically they were found to be composed of crowds of squamous epithelial cells, so that they resembled in structure the yellowish-white bodies developed in connection with the amnion of the Cow (p. 25).

Manis.—In *Manis*, as was pointed out by Dr Sharpey[1], the chorion was covered with fine reticulating villous ridges, interrupted here and there by round bald spots, giving it an

[1] Quoted in Huxley's *Elements of Comparative Anatomy*, p. 112, 1864; and with additional details in my Memoir on the Placentation of the Sloths in *Trans. Roy. Soc. Edinb.* 1873.

alveolar aspect, something like the inside of the human gall-bladder, but finer. A bare band free from villi ran longitudinally along the concavity of the chorion, and there was a corresponding bald space on the surface of the uterine mucous membrane. The ridges of the chorion started from the margins of the bald stripe and ran round the ovum. The mucous membrane of the uterus possessed a finely corrugated appearance on the surface, in correspondence with the ridges on the chorion, and was punctated with the orifices of numerous shallow crypts in which the villi had been lodged. Branched cylindriform glands were very abundant in the deeper layer of the mucosa, but their mode of opening on the surface could not be satisfactorily ascertained, owing to the condition of the specimen.

Camels.—The *Camelidæ*, unlike the ordinary Ruminant mammals, possess, as has for many years been recognised, a diffused placenta. I have recently, through the courtesy of Professor Flower, had the opportunity of examining a considerable part of the chorion of a Dromedary, preserved in the Museum of your College. The free surface of the chorion was thickly studded with short villi, but at one spot a bare patch about $1\frac{1}{2}$ inch long was seen; the relation of which to the wall of the uterus, owing to the absence of that organ, could not be ascertained. The villi were not arranged in tufts but arose singly from the chorion, having a somewhat constricted base, and expanding at the free end in a club-shaped manner. The villi varied in length from about the $\frac{1}{20}$th to $\frac{1}{12}$th inch. The larger proportion were unbranched, but some of the longer villi were divided into two or three short offshoots at the free end. The vessels of the chorion had been injected with size and vermilion, and a beautiful intra-villous network of capillaries was displayed. An extra-villous plexus of capillaries, not unlike that which I have described in the Mare, in *Orca* and in the Narwhal, was also seen. Although the gravid uterus itself was not in the museum for examination, yet there can be no doubt that its free surface must have been thickly studded with crypts for the reception of the villi.

I have examined the non-gravid uterus of the Alpaca (*Lama paca*) with reference to the arrangement of the utricular glands.

The mucous membrane was thicker than the muscular coat. The glands were numerous; the deeper ends were very tortuous and branched repeatedly, so that in vertical sections through the membrane they were many times divided. As they ascended to open by distinct mouths on the free surface they became much more straight, so that long portions of the tubes could be seen. The intervals between the glands occupied by the connective tissue was about equal to the transverse diameter of the glands. A columnar epithelium lined the glands and covered the free surface of the mucosa.

In the *Tragulidæ* also, as has been described and figured by M. A. Milne-Edwards in *Tragulus Stanleyanus*[1], the villi are not collected into cotyledons, but are uniformly diffused over the surface of the chorion.

In the non-gravid uterus of a *Tragulus javanicus*, which I examined, the glands were tortuous and branched in the deeper part of the mucosa, so that they were repeatedly divided in vertical sections through the mucous membrane. As they approached the free surface they became much straighter, so that long portions of gland-tubes could be seen passing to open on the surface. The glands were so close together that the interglandular tissue between any two glands was about as broad as the transverse diameter of the gland-tube. The epithelium in the glands and on the free surface of the membrane was columnar.

In the *Tapir*, as was shown by Sir Everard Home[2], the chorion is villous as in the mare. In *Tapirus Malayanus*, I am told by my friend Dr John Anderson, there is "a long bare area as in *Manis*, *Platanista* and *Orcella*, but proportionally of much greater size. The uterus also has the general characters of that organ in the gravid *Platanista*[3]." In the *Hippopotamus* M. H. Milne-Edwards has described[4] large villi disseminated over the whole surface of the chorion, except at the poles, where the membrane is smooth. Mr A. H. Garrod

---

[1] *Ann. des Sciences Naturelles*, p. 101, Vol. II. 1864.
[2] *Lectures on Comp. Anatomy*, v. p. 328, and Plate 27.
[3] The characters of the gravid uterus in the rare Cetacean genera, *Platanista* and *Orcella*, have been specially studied by Dr Anderson, and will be described by him in a memoir to be shortly published.
[4] *Leçons sur la Physiologie*, IX. p. 562, 1870.

has also seen[1] the uniformly villous covering of the chorion in the placenta of this animal.

Lemurs.—I have already stated (p. 20) that M. Alphonse Milne-Edwards has described the Lemurs as possessing a bell-shaped placenta. Since that was written I have had the opportunity of examining specimens of gravid uteri in a well-advanced state of development in *Propithecus diadema* and *Lemur rufipes*[2]. In all these specimens the mucous membrane in the anterior and middle thirds of the uterus was thrown into numerous shallow convoluted folds separated by intermediate sulci, constituting "une multitude d'aufractuosités irrégulières" as described by M. Milne-Edwards. Multitudes of crypts were situated on the summits and sides of the folds, as well as at the bottom of the intermediate sulci. The crypts in their general form and arrangement closely corresponded to the uterine crypts in *Orca* and in the Narwhal, though no glands were seen to open into them. They were lined by a columnar epithelium, beneath which was a very compact capillary plexus, similar in arrangement to that seen in *Orca* and in the Narwhal. Scattered amidst the crypt-covered folds were irregularly elongated smooth depressed areas, the largest of which did not exceed half an inch in their long diameter.

Utricular glands, somewhat tortuous in their course and occasionally branching, were seen in the deeper layer of the mucosa. They converged to the smooth areas just described, ran for a little distance beneath the membrane, and then opened, usually by an obliquely directed mouth, on the surface of the smooth area; as many as ten, twenty, or even a greater number of gland orifices, being found in a single area. These areas corresponded therefore to the smooth depressed spots on the surface of the gravid mucosa of the pig, but in the Lemurs each area had a much larger number of glands opening in it. It is clear therefore that in the Lemurs, as in the pig, the crypts are to be regarded as situated in the interglandular part of the surface of the mucous membrane.

---

[1] *Proc. Zool. Soc.* Nov. 19, 1872.

[2] I am indebted for these specimens to the kindness of Dr Andrew Davidson, of Antananarivo, Madagascar, and have communicated to the Royal Society of London a detailed description of their characters.

The chorion was elongated in form, but gave off from the depending part, lying near the os uteri, a short diverticulum, which passed into the cavity of the uterine cornu in which the fœtus was not situated; so that the chorion occupied not only the body, but both horns of the uterus. A large part of the surface of the chorion, more especially its anterior and middle thirds, was traversed by ridges from which numerous villi projected. The ridges fitted into the sulci between the folds of the mucosa, and the villi occupied the crypts. By gentle traction the chorion could be raised from the mucosa and the villi drawn out of the crypts, as in the Mare, Narwhal, or other animals with a diffused placenta. The posterior third of the chorion was smooth and without villi, and the corresponding area of the mucosa, surrounding the os uteri, was also smooth and without crypts. Smooth patches on the chorion also corresponded to the irregularly elongated, depressed, smooth areas on the surface of the mucosa. In *Lemur rufipes* a smooth surface was seen on the mucosa extending from the orifice of the Fallopian tube into the fundus of the horn in which the fœtus was lodged, and a smooth patch was situated on the pole of the chorion which corresponded to that surface; but in the *Propithecus* these surfaces were on the one hand covered with villi, on the other with crypts. In *Lemur rufipes* the chorion prolonged into the non-gravid horn was mostly non-villous, and the mucous lining was almost free from crypts.

The arrangement and structure of both chorion and uterine mucosa in the *Lemurida* prove these animals to possess a diffused placenta, and as, not only the surface of the chorion near the os uteri, but both its poles may be devoid of villi, the comparison which M. A. Milne-Edwards has made between its form and that of a bell is not generally applicable. The statement, which has frequently been made, that the Lemurs have a disc-shaped placenta, is therefore quite without foundation.

STRUCTURE OF THE POLYCOTYLEDONARY PLACENTA.

SINCE the time of Fabricius, who figured and described two and a half centuries ago the gravid uterus of the Sheep, the Polycotyledonary Placenta has been known to be the form of placenta in animals belonging to the order *Ruminantia.* It consists of a number of thick tuft-like masses of villi—the fœtal caruncles or cotyledons, which are lodged in crypts situated in an equal number of thick, spongy elevations of the uterine mucous membrane—the maternal cotyledons. The fœtal cotyledons are separated from each other by considerable areas of smooth chorion, and the maternal cotyledons have equally large areas of smooth mucous membrane between them. Each cotyledon is complete when the fœtal are lodged within the maternal cotyledons, and each when complete forms a miniature placenta.

The first indication of the formation of a maternal cotyledon, as has been pointed out by Ercolani[1], is an elevation of the uterine mucosa, so as to present an irregular undulating surface. As the development proceeds this irregularity increases, until well-defined depressions or crypts are formed. The walls of the crypts continue to grow in length, and the crypts are not only deepened, but smaller compartments branch off from them. In the course of time they assume the appearance of deep pits, subdivided into numerous crypt-like compartments, into which the villi of the chorion closely fit.

The shape of the fully-formed cotyledons and the disposition of the pits vary in different Ruminants. I shall describe in detail what I have seen in the well-advanced gravid uteri of both the Sheep and Cow.

Sheep.—In the Sheep the maternal cotyledons, as has frequently been pointed out by other observers, projected as cup-shaped mounds from the uterine wall. They were covered on the outer convex surface by the uterine mucosa, which was

---

[1] *Mém. dell' Accad. delle Scienze di Bologna*, 1870, Plate I. and 1873, Plate II.

smooth on its free surface, and was prolonged as far as the free
inverted edge of the cup.　The inner substance of the cotyledon was composed of a soft, spongy material, containing numerous pits, which extended almost vertically, and divided as they
passed deeper into its substance into smaller crypt-like compartments, which radiated towards the outer wall of the cotyledon, without diverging much from each other.　The pits were
lined by well-marked cells, most of which were irregular in

Fig. 11.

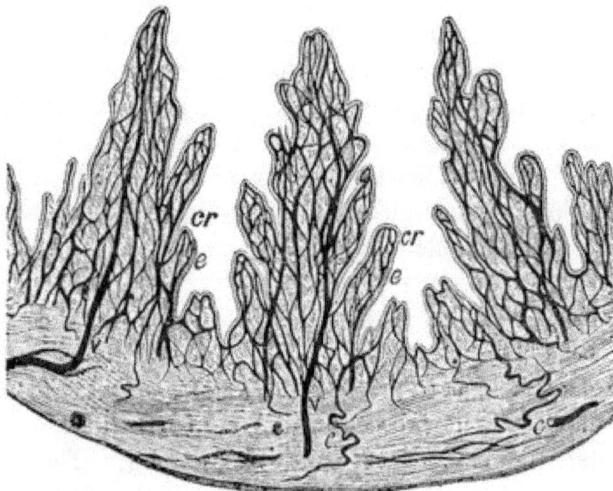

Semi-diagrammatic vertical section through a portion of a maternal cotyledon
of a Sheep. *cr. cr.* pit-like crypts, with *e. e.* the epithelial lining. *v. v.*
the veins, and *c. c.* the curling arteries of the sub-epithelial connective tissue.

shape, polygonal, ovoid, or even somewhat caudate, and of considerable size, though some appeared like modified columnar
cells.　They consisted of granulated protoplasm, in which one,
two, or sometimes three, well-defined ovoid or elliptical nuclei
were imbedded, but without a cell-wall.　Not unfrequently the
outline of the individual cells was very indistinct, and they
seemed as if arranged as a layer of protoplasm studded with
nuclei.

　　The cells rested on a highly vascular sub-epithelial connective tissue, which formed the proper wall of the pits.　The
mucous membrane investing the cotyledon was continuous at

the mouth of the cup with the walls of the pits in the spongy tissue, so that the cells lining the pits were in the same morphological plane as the epithelium covering the mucosa. The cotyledons were highly vascular. Some of the arteries in the sub-cotyledonary connective tissue were corkscrew-like; and in the deeper part of the cotyledon itself I have seen tortuous vessels. The greater number of the vessels within the cotyledon passed, however, vertically towards the surface, lying in the connective-tissue walls of the pits; and branching repeatedly, as a rule, in a dichotomous manner, prior to forming a compact maternal capillary plexus, but not dilating into maternal blood-sinuses.

The mucous membrane of the uterus between the cotyledons was smooth on its free surface, and contained numerous tortuous, branched, tubular glands. Some of these extended almost vertically to the surface, and could be seen in almost their entire length in vertical sections—others ran more obliquely, and, owing to their tortuosity, were repeatedly divided in vertical sections. The mouths of the glands could readily be seen with a pocket-lens opening on the surface, the orifice being partially surrounded by a minute elevation of the mucosa. In the mucosa around the base of the cotyledons a ring-like series of gland-openings was seen. In the mucosa covering the cotyledons, glands were also present, but their orifices were much stretched, as if by the pressure due to the great growth of the subjacent spongy tissue of the cotyledon. The sub-epithelial connective tissue, in which the glands lay, was not by any means so vascular as that which formed the walls of the pits within the cotyledons. In some sections through the cotyledons and adjacent mucosa no glands were to be seen in the connective tissue intervening between the cotyledon and muscular wall, but they were collected in considerable numbers around the cotyledon, as if pushed outwards by its rapid growth. In other sections, however, tubular glands were seen in the sub-cotyledonary connective tissue; but they seemed to be the deep ends of branching glands, the stems of which had inclined obliquely, so as to open on the surface of the mucous membrane covering the cotyledon. None of these subjacent glands, or those situated on the surface of the cotyledon, were seen to open into, or in any way to communicate with, the pits within the cotyledon itself.

The fœtal cotyledons consisted of numerous villi, which collectively formed a ball-like mass, called caruncle by the veterinary anatomist, and occupying the concavity of the maternal cotyledon. Each villus consisted of a main stem, which gave off a tuft or cluster of spatulate branches. The villi entered the maternal pits and branched along with them, so that every compartment was occupied by a branch of the villus; but there was necessarily no great divergence of these branches from the main stem. At their deeper end these spatulate branches gave off slender terminal offshoots. The villi were formed of gelatinous connective tissue, in which very distinct fusiform and stellate corpuscles were arranged in an anasto-mosing network. At the periphery of the villus was a layer of flattened cells, with small but distinct nuclei arranged so as to form an epithelial-like investment. The umbilical vessels rami-fied within the villi and formed networks of capillaries. The villi were in close contact with the epithelial cells lining the maternal pits. Owing to the inversion of the free edge of the maternal cotyledon and the radiated arrangement of the pits, with their contained villi, it was impossible to disengage the maternal and fœtal cotyledons from each other without drawing away with the fœtal villi portions of the maternal cotyledon. I invariably found that, in drawing the fœtal villi out of their compartments, flakes of epithelial cells accompanied them, which showed how readily this element of the maternal issue is shed. During parturition, however, when the parts are relaxed, the disengagement of the two structures can necessarily be more easily accomplished.

Cow.—In the Cow the maternal cotyledons, as is well known, differed in form from those in the sheep. They were fungiform or umbrella-shaped, and were connected to the uterine wall by a broad peduncle, around which the uterine mucosa was prolonged as far as the border of the umbrella. The whole convex surface of the cotyledon was riddled with pits, which passed vertically into its spongy substance, and divided into smaller compartments in the deeper part of the cotyledon. Projecting from the wall of each pit were delicate bands, visible to the naked eye, arranged as a rule in a vertical direction, and in the intervals between these bands the wall was perforated by numerous orifices, easily seen with a pocket-lens, which were the mouths of depressions or crypts in the wall of the pit, some lying almost at right angles, others obliquely to the wall of the pit itself. The pits, with their numerous crypts, were lined by cells, similar in character to those of the sheep. But I should state that a larger proportion of these cells had preserved the columnar form of the epithelium of the non-gravid uterine mucosa. They rested on a highly vascular connective tissue, in which the maternal capillaries formed a compact network.

The surface of the uterine mucosa between the cotyledons was perforated with the mouths of the tubular, branched, utricular glands. Each gland as a rule opened alone, but occasionally two lay together, a narrow band of mucosa intervening. Sometimes the mouth was a distinct funnel leading vertically into the gland-stem, but more usually the gland-stem was directed obliquely to the surface, so that in vertical sections through the membrane the glands were frequently cut through and divided. Glands were also present in the mucous membrane covering the peduncle of the cotyledon, but instead of communicating with the pits they opened on the free surface of the mucosa.

The fœtal cotyledons were concave and adapted to the fungiform maternal cotyledons, the pits of which were occupied by their villi. The stems of the villi were comparatively large, and studded with multitudes of minute tufts, which, arising obliquely or almost at right angles to the main stem, entered and occupied the crypts. The minute villi forming these tufts were so slender and filiform that each terminal offshoot contained only a single capillary loop. The villi were in contact with the epithelium-cells, and in drawing them out of the pits, more especially in drawing the tufts out of the crypts, multitudes of cells of the lining epithelium came away with them. From the differences in shape of the maternal cotyledon in the Cow and in the Sheep, there is not the same difficulty in unlocking the fœtal from the maternal placenta in the former animal as in the latter.

In both the Cow and Sheep the branches of the umbilical

artery and vein, which passed to and from the cotyledons, ramified in the smooth inter-cotyledonary part of the chorion. Numerous small branches arose from them, which did not enter the cotyledons, but dividing in an arborescent manner ended in a compact capillary plexus which formed a layer immediately beneath the smooth epithelial-covered surface of the chorion. This plexus was nearer the surface than the arteries and veins with which it was continuous. Slight variations in the closeness of the network were seen; in many places the capillaries were as small and the mesh-work as fine as in the walls of the pulmonary air-sacs, though elsewhere they were a little larger and the mesh-work more open. The plexus obviously represented on a large scale the extra-villous plexus already described in the diffused form of placenta. In both animals the highly vascular surface of the chorion presented a striking contrast to the non-vascular yellow-coloured *diverticula allantoidis* which projected through and beyond the poles.

Von Baer has described in his *Untersuchungen* (p. 18) the highly vascular character of the inter-cotyledonary part of the chorion in the Cow and Sheep, and states that a successful injection in the sheep forms one of the most beautiful anatomical objects with which he is acquainted. He further points out that these Ruminants possess in the later period of pregnancy a very large number of smaller "cotyledons" of a peculiar kind. In the Sheep they are so small and transparent as to be recognised with difficulty, though when injected they are more distinct. In the Cow they are about a line in diameter, and more easily recognised. They consist, he says, of collections of folds of the chorion, which are more vascular than the surrounding surface. E. H. Weber has also seen these structures, and speaks[1] of them as consisting of "cells" separated from each other by partitions, and having branches of the umbilical vessels distributed to them. He regards them as receptacles for the secretion of the utricular glands[2]. But

---

[1] *Hildebrandt's Anatomie,* iv. 506, 1832.

[2] Von Baer, who, as has already been stated (note, p. 37), at one time regarded the structures, now known to be the utricular glands, as lymphatic vessels, believed the collections of folds to be about as numerous as the terminations of the so-called "lymphatics" on the surface of the uterine mucous membrane, and that these folds corresponded with the star-like spots on the surface of the chorion of the pig.

little attention seems to have been paid by subsequent writers to these observations of von Baer and Weber, though Franck in his recently published treatise on Veterinary Anatomy[1] probably refers to the same structures when he says:

"Besides the proper cotyledons, numbers of small villi of 1 mm. in diameter, covered by epithelium, are found on the outer surface of the chorion in the first half of pregnancy. They do not completely disappear, and in the last third of pregnancy one finds in addition to the proper fœtal villi numbers of small rounded groups of similar papillæ, which lie loosely in apposition with the uterine mucosa, and in some measure represent a second series of small fœtal villi. At times some are more developed and become connected with the mucosa in places where no cotyledonary villi are formed."

I have examined the chorion of a cow, the fœtus in which measured 10 inches from the nose to the root of the tail, with the especial object of determining the nature of these structures. The free surface of the inter-cotyledonary part of the chorion was mottled with faint yellow spots, the largest of which were about 1 line in diameter, and between 25 and 30 spots were on an average in each square inch. The chorion at each spot was slightly thicker and more opaque than in the surrounding surface. In non-injected specimens their structure was not very distinct, but where the inter-cotyledonary capillaries had been injected their characters could without difficulty be ascertained. The simplest form was a shallow pouch, or pocket-like depression, opening on the outer surface of the chorion. Sometimes a pocket lay singly, but more usually two, three, or more, were closely grouped together, and in one case I saw as many as twelve placed side by side. Not unfrequently a pocket had a circular outline, but it was often elongated, and occasionally had a horse-shoe form. The walls of the pockets were formed of the chorion, and a compact capillary plexus was distributed in them, and frequently had a whorl-like arrangement around the open mouth of the pocket. The pockets were lined by an epithelium continuous with the squamous epithelium covering the plane surface of the chorion, and their epithelial lining was not so easily washed off as that on the plane surface.

---

[1] *Handbuch der Anatomie der Hausthiere*, p. 1017. Stuttgart, 1871.

Under a magnifying power of 320 diameters numerous yellow granules were seen in the epithelial lining, which seemed occasioned by a fatty degeneration of the nuclei of many of the cells. From this description it is clear that these spots cannot be regarded either as cotyledons, or as villi, in the strict sense in which these terms are now employed by anatomists. It was not possible to say precisely what part of the uterine mucosa they were in direct apposition with; but from the general correspondence in numbers between the spots and the mouths of the utricular glands it is most probable that they were opposite the gland orifices. Hence I agree with Weber in regarding them as receptacles for the secretion of these glands.

Deer.—In the Red-deer the cotyledons do not exceed twelve in number, and vary in diameter from ¼ to 1 inch. They are not scattered generally over the surface of the uterus and chorion, but are arranged in linear series along the lesser curvature of each cornu. The uterine cotyledons are not pedunculated as in the Cow, though their free surface is convex. The villi of the chorion formed tufts, the branches of which resembled in shape those of the Sheep, though somewhat longer than in that animal.

The form and to some extent the structure of the cotyledons in the Roe-deer have been described by E. H. Weber[1] and Bischoff[2]. The number of cotyledons, which in the Cow and Sheep amounts to from 60 to 100, does not in the Roe-deer exceed 5 or 6. The long vascular villi of the fœtal cotyledons fit into the tubular depressions in the maternal cotyledons, in the walls of which a capillary plexus is distributed. Bischoff has satisfied himself that no utricular glands exist in the maternal cotyledons. He figures numerous ramifications of the umbilical vessels in the inter-cotyledonary part of the chorion.

Giraffe.—Prof. Owen has given a description with some beautiful figures of the chorion of a Giraffe[3] shed during parturition; and has pointed out that numerous cotyledons, some of which were 4 inches in diameter, and of an oval or reniform

---

[1] *Hildebrandt's Anatomie*, IV. 506, 1832.
[2] *Entwicklungsgeschichte des Rehes*, Giessen, 1854.
[3] *Trans. Zool. Soc.* Vol. III.

shape, were arranged in longitudinal rows. I have examined microscopically the villi from the larger cotyledons of this specimen as preserved in your Museum, and have found them to possess some variations in form. Some were filiform and almost cylindrical, others broader and more flattened. Some were unbranched except at the free end, where they gave origin to two or three short bud-like offshoots: others were much more deeply cleft, but none could be said to have an arborescent form. The cotyledons were very vascular, and each villus contained a compact capillary network.

But, further, Owen has described (*Op. cit.*) numerous smaller cotyledons

"of irregular form and unequal dimensions, developed from the external surface of the chorion in the interspace of the normal cotyledons : these smaller ones varied in diameter from two inches to two lines : their component villi were proportionately short, and in the smallest ones simple and unbranched ; so that the parts of the chorion where these were thickly scattered, presented a structure approaching to that of the non-placentiferous chorion of the Camel and certain Pachyderms."

In addition to the smaller cotyledons described by Owen, I have seen, under a magnifying power of 40 diameters, short club-shaped villi springing from the plane surface of the chorion, either singly, or in clusters of 2, 3 to 6 villi, or in linear rows, containing as many as 20 short, simple, clavate villi closely resembling in form the villi of the Camel, so that the Giraffe in this particular approximates more to the diffused placenta than do the Cow, Sheep and Deer.

But the chorion of the Giraffe presents other peculiarities. The free surface of the inter-cotyledonary chorion was mottled with irregularly-shaped spots, varying in diameter from half an inch to one line ; so numerous indeed were they, in the portion of chorion I had for examination, that they occupied the larger proportion of the surface, and left only very narrow lanes of unspotted chorion between. The membrane was slightly thickened at each spot, but was not elevated as with the smaller cotyledons of Owen. Under a simple lens the free surface of each spot was minutely pitted, and when magnified 40 diameters the pits were seen to be pocket-like depressions similar in form, arrangement and structure to those that I have described in the corresponding part of the chorion of the Cow. Owing to the spots being so much bigger than in the Cow, the number of pockets in a spot was much greater, 100 and upwards being not unfrequently grouped together. The pockets were in some places separated from each other by well-defined septa just as in the Cow, but in other members of the group the septa were split up into villous processes, and adjacent pockets communicated with each other. The short villi which sprang from the plane surface of the chorion were usually found in close proximity to a group of pockets. The chorion had been minutely injected with size and vermilion, and a compact plexus of capillaries was seen immediately beneath its free surface, as in the Sheep and Cow ; they were also distributed in the walls of the pockets, in the villous subdivisions of the septa and in the scattered villi. Fragments of the epithelial lining could be seen in some of the pockets, though for the most part it had been removed. There can, I think, be no question that the pockets in the Giraffe fulfil the same office as the corresponding structures in the Cow.

Although the uterine mucosa of the gravid Giraffe has not apparently been examined, there can be no doubt that maternal cotyledons containing pits for the reception of the fœtal villi must exist, and Owen has shown that, even in the non-gravid state, elevations are to be seen in the uterine wall, which correspond in position to the future cotyledons. Utricular glands also without doubt exist, and it is probable from the number of spots composed of pockets in the chorion that they occur in large numbers.

It is necessary that we should now consider whether the pits and crypts in the maternal cotyledons of the ruminant placenta, in which the villi of the chorion are lodged, are merely the greatly-enlarged mouths of the utricular glands of the mucosa, or are structures specially formed during pregnancy, by great hypertrophy and folding of the inter-glandular part of the mucous membrane, as in the diffused form of placenta. Prof. Spiegelberg[1] was of opinion, from some observations which

---

[1] *Henle and Pfeuffer's Zeitschrift,* xxi., quoted by Ercolani, p. 17 of the French translation of his Memoir.

he had made, that they were only remarkable dilatations of the utricular glands, and Bischoff was at one time disposed to regard them as the largely-developed glands of the uterus. Subsequently however Bischoff figured, in the uterus of the Roedeer[1], the utricular glands ascending to open on the surface of the uterus, not in the cotyledons, but around its circumference. Eschricht however had previously stated that in the Cow the glands open, not into the cotyledons, but on the intermediate surface of the uterus. Ercolani also has figured and described[2] both in the Sheep and Cow the glands as situated around the cotyledons, and not communicating with the cavities within them. In the description which I have given of the cotyledons in the Sheep and Cow, I have stated that I was unable to detect any communications between the glands and crypts : the glands indeed appeared as if they had been pressed to the periphery of the cotyledons by the great development of its spongy substance. Hence it would appear that in the polycotyledonary, as in the diffused placenta, the crypts in which the fœtal villi are lodged are not produced by an enlargement and dilatation of the tubular glands of the mucosa; but are new structures formed, during pregnancy, by a great hypertrophy and folding of the inter-glandular part of the mucous membrane.

[1] *Entwicklungsgeschichte des Rehes*, Plate VIII. 1854.
[2] *Memoir of* 1873, Plate 2.

STRUCTURE OF THE ZONARY PLACENTA.

THE Zonary or Annular placenta forms a belt surrounding
the equatorial part of the chorion and contiguous portion of the
uterine mucous membrane. It is found in its best-known
form in the *Carnivora* and *Pinnepedia*, though it occurs also in
Hyrax and in the Elephant.

But in some of the *Carnivora* modifications in the form of
the zone are met with. Thus Daubenton pointed out[1] that in
the Ferret (*Mustela furo*) the zonary arrangement is so far
modified that two portions of the zone are thicker than the rest,
so as to look like two lobes, but an intermediate thinner tract
unites these lobes together. Bischoff has described in the Stone-
marten (*Mustela foina*), and Pine-marten (*Mustela martes*)[2], on
the side corresponding to the free border of the uterus, a gap in
the continuity of the ring-like band of the zonary placenta, in
which gap the chorion formed a pouch. The villi bounding
this gap were strongly developed, and like the outer surface of
the pouch itself possessed an epithelium, coloured of a rich
reddish yellow, in which the pigment was partly in the form of
granules, partly of rhombohedral crystals. The pouch contained effused blood, the liquid part of which was intensely
coloured and contained hæmatoidin granules and crystals. In
the Otter (*Lutra vulgaris*) a similar pouch has been described[3]
by Bischoff, but larger and containing more free blood. The
same anatomist has also pointed out[4] that in the Weasel (*Mustela
vulgaris*) the placenta does not form a complete zone, but is
divided into two lateral halves by two gaps, the one opposite
the free border of the uterus, the other opposite its mesenterial
attachment. The chorion at the space opposite the free border
was not quite smooth, but possessed some tolerably long villi,
and had a reddish-yellow colour on the surface. Between this
part of the ovum and the uterus some extravasated blood was
also found.

---

[1] Buffon's *Histoire Naturelle*, VII. Pl. 27.
[2] *Sitzbericht. Akad. Wissensch. München*, 13 May, 1865, p. 339.
[3] *Idem*, p. 213, 1865.
[4] *Idem*, p. 343, 1865.

It has long been known that the free borders of the placenta in the Bitch and Cat are coloured of a deep green, not unlike the colour of bile, and a beautiful drawing of this appearance in the Bitch has been published by von Baer in his *Untersuchungen*. Bischoff, who examined the coloured border microscopically, found[1] it to contain long pointed crystals, soluble in water; a beautiful green pigment in irregular granules; a quantity of small rounded globules, also soluble in water; larger feebly-granulated cells, with a round or somewhat elongated nucleus; a brown mass and a few large fat-cells. Barruel[2] had previously regarded the green colouring material as like that of the bile, and H. Meckel in a more recent paper had applied to it the name of hæmato-chlorine[3]. I have seen the placenta of the Fox with a similar green coloration at the borders of the zone.

The gravid uterus of the Dog, Cat, and other pluriparous *Carnivora* possesses a moniliform appearance. Each dilatation is a compartment of the uterus containing an embryo, with its membranes, and between adjacent compartments the uterine cavity forms a narrow tube. If one of these compartments be opened, in a well-advanced stage of development of the embryo, the chorion will be seen to be smooth and bare of villi, except in about its middle third, where the villi are arranged as a zonular band around the transverse circumference of the ovum. The uterine mucosa possesses a similar zone closely blended with the zonular band of the chorion. The mucous membrane on each side of the zone is smooth and vascular: it lies in apposition with the smooth part of the chorion, but has no attachment to it. Where the zonary and smooth parts of the mucosa are continuous with each other a narrow strip of mucous membrane is reflected on the margin of the zonular band of the chorion, and forms a rudimentary decidua reflexa. In the true *Carnivora* the decidua reflexa is so very narrow that it has often been overlooked; but in the Grey Seal, where the placenta is large, the reflexa is from ¾ to 1¼ inch broad. As the zonary placenta is much more complex in structure than either the

[1] *Entwicklungsgeschichte des Hunde-Eies*, p. 106.
[2] *Ann. des Sciences Nat.* xix. 1830.
[3] *Deutsche Klinik*, 1852, p. 466.

diffused or polycotyledonary forms, it is necessary, to understand
the relations of its constituent parts, that it should be examined
in different stages of development.   I shall describe what I have
seen in the domestic Cat.   (Pl. I. II., figs. 1, 2, 3, 9, 10.)

Cat.—In the earliest impregnated Cat's uterus, which I have
examined, the compartments were ovoid, and the long diameter
of each, measured along the arc, did not exceed $\frac{8}{10}$th inch.
When a compartment was opened the chorion readily separated
from the mucous lining.   At each pole of the compartment an
area of mucosa $\frac{1}{10}$th inch in its long diameter was smooth; but
the rest of the membrane was hypertrophied, spongy, swollen,
and elevated above the smooth polar portions, and formed the
placental area.   The placental area was marked on its surface by
an extremely delicate reticulation, many of the strands of which
had a sinuous direction.   It was thickly studded with minute
orifices barely visible to the naked eye, but easily seen with a
pocket-lens.   These orifices were the mouths of the pits or
crypts in which the villi of the chorion had been lodged.   A few
of these openings were two or three times larger than the rest.
The appearance which I saw in the Cat is evidently similar to
the "cells" figured by Dr Sharpey in the Bitch (Fig. 211)[1], and
by Bischoff in the same animal (Fig. 48, A)[2], though, as will be
seen further on, I interpret its mode of production in a different
manner from those anatomists.   The crypts passed vertically into
the spongy substance, and when vertical sections were made
through it, they were seen to be separated from each other by
trabeculæ; the chief beams of which lay vertically, and when
they reached the free surface formed the strands of the reticulum
already described.   The vertical trabeculæ were connected together by others directed obliquely or in a sinuous manner, and
these lateral connections were especially seen about midway in
their length.   Hence not only on the surface, but when horizontal sections were made through the placental area, a reticulated arrangement was seen, and the crypts constituted the
interstices of the reticulum.   As these trabeculæ were formed
of the thickened mucous membrane of the placental area, they

---

[1] Baly's Translation of *Müller's Physiology*, note p. 1576.
[2] *Entwicklungsgeschichte des Hunde-Eies*, 1845.

were necessarily composed of the somewhat modified tissues of that membrane. On the surface was a definite layer of epithelium, the cells of which were short columns, with distinct, circular or ovoid, brightly-refracting nuclei. These cells rested on a delicate sub-epithelial connective tissue in which the maternal capillaries ramified.

The trabeculæ and the sub-mucous connective tissue were carefully examined with the object of ascertaining their relations to the tubular glands. In vertical sections the glands were distinctly seen, transversely or obliquely divided, lying in a definite layer of connective tissue situated deeper than the crypts. Sometimes the divided glands were separated, by comparatively broad bands of connective tissue, from the crypts and trabecular structure, but in other places they were immediately subjacent. They were lined by a well-defined columnar epithelial layer. I examined many sections with the object of ascertaining if the stems of the glands opened into the crypts or passed along the trabeculæ to open on the free surface of the mucosa, but owing to the complexity of the arrangement due to the formation of the numerous crypts which I have described, I did not succeed in tracing them to their orifices.

As it was important however to ascertain if the crypts equalled in number in a given area the glands of the mucosa in the same area, or if the crypts much exceeded in number the glands, I submitted different parts of the mucosa of the gravid uterus of this cat to microscopic examination, and compared the appearances seen with those presented by the mucosa of the non-impregnated uterus. In the non-gravid cat the stems of the glands were almost perpendicular to the free surface of the mucosa. They were so tortuous at their deeper ends as to be repeatedly cut across in a vertical section through the membrane. The interglandular connective tissue, containing numerous corpuscles, formed well-marked septa between the glands. Vertical sections made through the mucosa lining the constrictions between the compartments of the uterus of this gravid cat showed the tubular glands to be on the average ¼th wider than in the non-gravid condition, the interglandular connective tissue was much smaller in quantity, so that the glands were more closely crowded together; but in the placental area of the mucosa of the same cat the interglandular tissue was greatly increased in quantity, so that the glands were much further apart. The glands themselves were, as in the non-placental area, dilated, but the number of glands seen in the sections did not nearly equal the number of crypts.

In a cat's ovum, which had reached a somewhat more advanced stage of development, where the long diameter of the uterine compartment, measured along the arc, was $1\frac{1}{2}$ inch, I found that the villi of the chorion readily disengaged from the uterine crypts. By far the larger part of the chorion was still villous, not more than $\frac{2}{16}$ ths inch at each pole being smooth. The line of demarcation between the placental and non-placental polar areas of the mucosa was very distinct. The placental area, or the hypertrophied and spongy mucosa, possessed a reticulated appearance, the principal strands of which were sinuous, and gave off numerous collateral branching offshoots, which joined adjacent branches to form the walls of the numerous pits or crypts which opened on the surface. The strands and branches were larger and the pits and crypts were more dilated than in the younger ovum already described, and on looking down the larger pits, their subdivision into smaller crypts could be seen. The crypts were lined by an epithelium, numbers of the cells of which possessed a columnar form, though others were swollen and otherwise altered in shape, so as to be irregularly polygonal. The cell-protoplasm was granulated and the nucleus was distinct. The sub-epithelial connective tissue was vascular. When vertical sections were made through the placental area the more dilated size of the crypts and pits than in the younger specimen was distinctly recognised, being thus in conformity with the larger size of the chorionic villi. Between the deeper closed end of the crypts and the muscular coat was a definite layer in which portions of gland-tubes, lined by an epithelium, some of which were transversely, others obliquely divided, could be seen. The glands were dilated as in the younger specimen, and not so numerous as the crypts, neither could I obtain satisfactory evidence of the communication of the mouths of the glands with the crypts.

I am led therefore to the conclusion that the crypts formed in the early period of gestation in the placental area of the Cat are not due to a mere widening of the mouths of the tubular glands; but are produced, as in the pig and mare, by a great increase in the amount of the interglandular part of the mucosa, which becomes folded so as to form the crypt-like arrangement which I have just described. In this respect, therefore, my observations agree with those of Ercolani on the same animal[1]. The interpretation, therefore, which Ercolani and I have put on the appearances seen in the placental area of the cat in the early stage of gestation, differs from that given by Dr Sharpey on the appearance seen in the uterine mucosa of the bitch at a similar stage. As is so well known, Dr Sharpey held that the pits and "cells" (crypts) seen on the inner surface of the uterus, which receive the villi of the chorion, are the mouths of the utricular glands enlarged and widened. It is possible that in the cat, as in the *Orca*, the utricular glands may open into some of the crypts, so as to seem to justify the inference that they were formed by a widening of the mouths of the pre-existing glands. But this interpretation obviously cannot be given of the formation of those crypts which are interglandular in position. Hence it seems to be more in conformity with the structural arrangements of the organ to conclude, that the crypts which arise in the uterine mucosa during pregnancy are new formations, produced by a great hypertrophy and folding of the surface of the mucous membrane.

When the ovum of a cat, which had completed about one-half the period of gestation, was examined, a most important advance in placental formation was observed. The zonary villous band on the chorion was restricted to its middle third, and an equally large smooth surface was found at each pole. The zone on the chorion was now so completely interlocked with the corresponding zone in the uterine mucosa, that the two surfaces could not be disengaged from each other. The placenta could only be separated from the uterus by rupturing the slender marginal band of decidua reflexa, and tearing through, or altogether pulling off, the placental area of the mucosa, which formed a well-defined decidua serotina.

---

[1] *Mem. dell' Accad. delle Scienze di Bologna*, 1870, Plates 2, 3, 4.

The villi of the chorion had the form of broad sinuous leaf-lets, which became attenuated at their uterine ends and gave off bud-like offsets from the free border. When vertical sections were made through the placenta the villi were seen to pass vertically through the organ up to its uterine aspect. The trabeculæ of maternal tissue, which formed the walls of the pits or crypts in which the villi were lodged, passed between the villi up to the chorion, and closely followed the sinuosities of the villi, so as to form an intimate investment for them, and in horizontal sections through the organ they were seen to be arranged as a series of laminæ, winding in a sinuous manner between the leaf-like villi. Between the placenta proper and the muscular coat was a well-defined layer of serotina, equal in thickness to the muscular coat itself. It was traversed by the numerous blood-vessels which passed into and out of the placenta, and which formed not unfrequent anastomoses with each other. The decidua serotina consisted not only of the vascular connective tissue, but of the epithelial cells of this part of the mucosa, which were similar in character to those described in the preceding stage of development. In thin sections, tubes were seen cut transversely or obliquely, and lined by an epithelium; they were about equal in diameter to the gland-tubes seen in the serotina in a less advanced stage of gestation, and were without doubt the dilated glands of this portion of the mucosa. It may here be stated, that in the non-placental area of the same uterus the tubular glands were distinctly seen separated from each other by comparatively wide intervals of interglandular tissue.

The trabeculæ and laminæ of maternal tissue, which were prolonged into the substance of the placenta between the villi, were continuous with the serotina and were invested by an epithelial layer, the cells of which were modified columns, like the cells of the decidua serotina. The blood-vessels of the serotina entered the laminæ and trabeculæ and ramified in them throughout the maternal part of the placenta. In the placenta of one of the embryos, where the maternal vessels were injected, they formed a network of capillaries of ordinary magnitude. In the other placentæ from the same uterus the maternal capillaries when injected with red gelatine were dilated to two or three times the size of the capillaries in the fœtal villi, and ascended almost vertically in the trabeculæ. Not unfrequently near the chorionic surface they dilated into sinus-like enlargements, which were crowded with blood-corpuscles. It is possible that these dilatations may, to some extent, have been due to the force employed in filling the maternal vessels with injection, but this will not, I think, account for the whole extent of the dilatation[1]. The vessels of the capillary network of the fœtal villi were injected with a blue colour and showed no dilatations; and the contrast between the two systems of vessels within the organ was well seen both in horizontal and vertical sections.

The placenta of a cat, shed in the ordinary course of parturition, was covered on its uterine surface by a layer of soft yellowish-white tissue, which was smooth and uniform in character, and was without any flocculent, ragged processes projecting from it. This layer consisted of that part of the serotina which remained attached to the placenta, came away with it at the time of birth, and was therefore deciduous. From it laminæ and trabeculæ passed into the substance of the placenta, which had a similar sinuous arrangement and relation to the fœtal villi as has just been described in the placenta at half time. Examined microscopically, the vascular connective tissue of the intra-placental prolongations of the serotina with their epithelial investment was recognised, but as it was not possible in a detached placenta to inject the maternal bloodvessels their disposition could not be made out. I examined thin sections through the deciduous or placental layer of the serotina for the presence of utricular glands. I saw indistinct appearances of tubes transversely or obliquely divided, which might be interpreted as tubular glands, but the aggregation of cells within and around them was so great that it was difficult to speak positively on this point. The chorionic system of fœtal blood-vessels was injected, and the leaf-like villi, with their remarkable compact capillary plexus, were readily seen. On examining with a pocket-lens the uterine surface of the deciduous layer of the serotina, many minute, rounded, scat-

[1] The dilatation of the maternal vessels in the feline placenta has also been referred to by Eschricht and other observers.

tered holes were seen in it, through each of which a terminal bud of a leaf-like villus projected so as to reach the
uterine surface of the placenta. These buds were often clavate
in form, and contained a capillary plexus, continuous with that
of the body of the villus. It is clear, therefore, that when the
placenta of the cat is shed at the time of parturition, a continuous layer of serotina, interrupted only by these minute
orifices, is shed along with it.

The presence of a layer investing the uterine surface of the
cat's placenta, analogous to the caducous layer of the human
placenta, was distinctly recognised by Eschricht; who also described the thin, perpendicular, flexuose laminæ of maternal
structure passing through the entire thickness of the organ and
investing the fœtal villi as if with sheaths[1]. Though Eschricht
was at first inclined to the view that the layer investing the
uterine surface of the placenta was nothing else than the
mucous tissue of the uterus, further consideration led him to
state that it altogether differed from that tunic. But he also
came to the conclusion that the mucous tunic was left entire on
the placental zone, exhibiting only torn and broken-off vessels.

There can be no doubt, from both its position and structure,
that this layer belongs to the mucosa of that part of the uterus
which corresponds to the placental zone, for it and the intra-
placental laminæ and trabeculæ are merely a more advanced
condition of the crypt-like modification of the mucosa, which I
have described in the earlier stages of placental formation in
this animal. Does this layer however represent the whole
thickness of the mucosa belonging to the placental zone? or is
it merely the superficial part of the mucous membrane? are
questions which may now be asked. In the uterus of the cat
killed in the mid period of gestation, I found, on peeling off
the placenta, that the serotina did not split into two layers,
the one a deciduous serotina attached to the placenta, the
other a non-deciduous serotina remaining connected to the
uterine wall, but that the whole thickness of the mucosa came
away with the placenta, leaving the muscular coat exposed;
moreover, the uterine face of the placenta presented a smooth

---

[1] De Organis, &c., pp. 14, 18.

surface similar in appearance to that exhibited by the organ when shed at the full time; and a similar separation also took place in the process of injecting the vessels of the gravid uterus.

From these specimens I was at one time inclined to think that the entire thickness of the mucosa in the placental area of the cat was shed along with the placenta during parturition. As I knew however that this important point could only be satisfactorily decided by an examination of a uterus immediately after the birth of the placentæ, I procured, through the kindness of my friend Dr Foulis, the uterus of a cat which had been killed five hours after having given birth to four kittens. The uterus was contracted, and the mucous lining was elevated into rugæ. Each placental area was a narrow zonular trench bounded at each margin of the zone by a fold of the mucosa. The surface of the non-placental part of the mucosa was unbroken and covered by epithelium. The surface of the placental zone was blood-stained, and with a number of shreds of membrane hanging from it, so that it had a torn and flocculent appearance. When thin flakes were removed from the surface of the placental zone, and examined microscopically, they were seen to consist of multitudes of free red blood-corpuscles, of very delicate fibres of connective tissue, intermingled with which were fusiform and lymph-like corpuscles, and here and there a patch of cells, evidently epithelium. A series of vertical sections was then made through the placental area and adjacent non-placental part of the mucosa, and examined with both low and high magnifying objectives. The free edge of the section in the non-placental area was covered by a well defined layer of columnar epithelium, deeper than which was a thick layer of sub-epithelial connective tissue, intervening between the epithelium and the muscular coat. Lying vertically in this connective tissue were numerous utricular glands, which opened on the free surface of the mucosa, and were lined by columnar epithelium. In the placental area itself the surface epithelium was absent, and the free edge of the section had not a smooth outline, but was irregular and with slender filaments of connective tissue projecting from it. The thickness of the connective-tissue layer on the surface of the muscular coat was

appreciably less (on the average about one-third) than in the
non-placental area. In this connective tissue sections through
utricular glands were seen. Some of these sections were transverse to the tube of the gland, others oblique, others almost
longitudinal. The epithelial lining of the glands was present,
and it is not unlikely that the occasional patch of cells found on
the surface of the placental area may have belonged to the
glands and not to the surface epithelium. In more than one of
the sections I saw in the placental area gland-structures, which
had not the form of cylindrical tubes, but were somewhat
irregularly dilated. Numerous blood-vessels which were the
vascular trunks going to the placenta were also seen plugged
with collections of blood-corpuscles.

From this description it will be seen that in the normal
separation of the placenta at the time of parturition, so complete a shedding of the mucosa in the placental zone did not
take place as was effected by artificially tearing off the placenta
in an earlier period of gestation. But from a comparison of the
placental and non-placental areas in this uterus, it is clear that
during parturition not only is the epithelium, but a portion of
the layer of sub-epithelial connective tissue, shed along with
the placenta as a deciduous serotina; whilst the deeper part of
the connective tissue, with the remains of the blood-vessels and
glands, persists as a covering for the muscular coat, and forms a
non-deciduous serotina.

A few words may now be said on the non-placental part of
the chorion in the Cat. Many anatomists have pointed out
that in the *Carnivora* slender branches of the umbilical vessels
pass as far as the poles of the chorion; but it has not sufficiently been recognised that they form immediately beneath
its free surface a compact capillary plexus, which, though
not quite so close perhaps as in the smooth parts of the
chorion of the Cow and Sheep, is yet so abundantly provided with capillaries as to give to this part of the chorion,
when the vessels are injected, a distinct colour. This well-
marked vascularity of the smooth chorion is seen not only in
the membrane about the mid-period of gestation, but in specimens shed at the normal period of parturition.

The outer surface of the smooth chorion of the cat possesses also another character worthy of notice. In specimens about the mid-period of gestation, the membrane was mottled with faint yellowish spots and lines, so as not to be perfectly translucent. When examined with a pocket lens the outer surface had a faintly corrugated appearance, as if slightly roughened with some extraneous material, which could easily be scraped off with a knife. When examined microscopically, this substance was seen to consist of cells which varied considerably in form. Some were flattened scales, others were more elongated, or even columnar, others again were rounded; and, in nearly all, the nucleus was relatively large and very distinct. The cells were probably produced by a proliferation of the epithelial cells normally coating the free surface of the chorion. The corresponding surface in the shed placenta of the cat was free from corrugations, but had a clouded mottled appearance, so as not to be uniformly translucent. Where the faint opacities were present large collections of very fine granules were visible, amidst which the outlines of nuclei and cells could be indistinctly seen, so that the opacity seemed due to collections of epithelial cells undergoing a minute granular (fatty) degeneration.

Bitch.—Though the placenta in the Bitch, as in the Cat, possesses the zonary form, yet its minute structure in these two animals presents sufficient differences to enable the anatomist readily to distinguish the one from the other. If the description and figures by Sharpey and Bischoff of the early stages of formation in the Bitch be compared with the corresponding stages in the Cat, a close resemblance is seen; but in the more advanced stages characteristic differences can be recognised.

In the Bitch, both at half and full time, when the placenta was stripped off the uterine zone, a distinct mucous membrane was left on the uterus, which was continuous at the margins of the zone with the narrow band of decidua reflexa and through it, with the mucosa covering the non-placental area. This zonary mucous membrane was subdivided into numerous irregular polygonal pits or trenches, bounded by folds of the mucous membrane. These folds had a ragged, flocculent appearance. The membrane was very vascular, and at the ragged edges of the fold numerous torn blood-vessels were seen. When examined microscopically the free surface not only of the pits and trenches, but of the folds, was seen to be covered by a layer of cells—the epithelium of the mucous membrane—which rested on the vascular sub-epithelial connective tissue. When this epithelium was looked at from the surface, a pattern of polygonal cells was seen like the free ends of columnar epithelium; but the cells were bigger than one usually finds this form of epithelium to be, and had, more especially in the uterus at full time, a distinct yellow colour, as if the cells were undergoing fatty degeneration. When the cells were scraped off, so as to be seen in profile, their columnar form was easily recognised. As this mucous membrane was not detached from the uterus along with the placenta, it is to be regarded as a non-deciduous serotina.

The uterine surface of the placenta also had a ragged appearance, for the numerous folds of the mucous membrane had entered the placenta, and, when it was stripped off, their torn ends were seen on its outer surface, but the flocculent appearance was still further increased by the free ends of the chorionic villi, which reached the surface. The prolongations of the mucous folds entered the placenta at a multitude of points in the interspaces between the villi, and as they ascended to the chorion they branched repeatedly, so as to give investments to the branches of the villi of the chorion. These intra-placental prolongations of the mucosa consisted of sub-epithelial connective tissue, in which the maternal vessels ramified, and of an epithelium composed partly of columnar cells, and partly of cells the regular columnar form of which had been modified into irregular polygons. These cells were larger and more distinct than the cells on the corresponding structures in the cat, and their protoplasm was so very granular as in many cases to obscure the nucleus. These prolongations of maternal tissue constituted a deciduous serotina. The shed placenta of the Bitch, whilst possessing in its substance numerous prolongations of maternal tissue not unlike those previously described in the Cat, yet differs from the latter animal, as has also been pointed out by Prof. Rolleston[1], in the absence of a continuous layer of deciduous serotina on its uterine aspect.

[1] *Trans. Zool. Soc.* v. 1863.

The chorionic villi in the bitch were not so sinuous and leaf-like as in the cat. They were more subdivided, and branched, so as to terminate in short villous tufts. Branches of the umbilical arteries ended within the villi in a compact capillary plexus. The villi were in close contact with the epithelial cells investing the intra-placental prolongations of the mucous membrane.

The non-placental areas of the chorion, as in the cat, contained ramifications of the umbilical vessels ending in a capillary plexus. The membrane was however uniformly translucent, and in none of my specimens did I see collections of epithelial cells such as I have described in the cat.

I may now relate some observations which I have made on the glands in the non-gravid uterine mucous membrane of the Bitch. It is well known that two kinds of glands were described by Dr Sharpey[1] in the uterine mucous membrane of this animal, viz. short, simple, unbranched tubes, and compound tubes having a long duct dividing into convoluted branches, both kinds opening close together on the surface of the mucosa. These observations were supported by Weber and Bischoff, and generally accepted by anatomists and physiologists; but Prof. Ercolani of Bologna, in his first memoir on the Structure of the Placenta[2], stated his inability to distinguish more than one kind of gland, and concluded that only the long tubular glands were present. I have felt it necessary therefore carefully to examine the uterine mucous membrane of the unimpregnated bitch with reference to this question. On a surface view the mouths of the glands could be distinctly seen closely crowded together, as is so well represented in Dr Sharpey's figure (fig. 209), and in Bischoff's memoir (*Entwicklungsgeschichte des Hunde-Eies*, Plate XIV. Fig. 47). When horizontal sections were made through the membrane near its surface the glands were seen to be transversely divided, and so closely set together that the interval between any two adjacent glands was in some cases not equal to, in other cases about equal to, the transverse diameter of a gland-tube; further, all the gland-tubes in any given transverse

---

[1] Baly's Translation of *Müller's Physiology*, Note, p. 1576.
[2] *Mémoire sur les Glandes Utriculaires de l'Uterus*, p. 22, French Translation, Algiers, 1869.

section exhibited the same structural characters. The inter-glandular connective tissue contained not only the usual fusiform corpuscles, but round cells like those of lymph or the white blood-corpuscles. When vertical sections through the membrane were examined, long compound tubular glands were readily seen passing into the deeper part of the mucosa, and between these, short and simple tubes were also recognised, so that, under low magnifying powers, at first sight these sections seemed to confirm the observations of Sharpey, Bischoff and Weber, which were made under magnifying powers of 10 and 12 diameters. When magnified more highly these apparently short simple glands were seen to vary considerably in length, some dipping for only a short distance from the surface of the mucosa, others for a greater distance, and exhibiting indeed every gradation in length up to the branched tubular glands themselves. But in the connective tissue, immediately deeper than the short glands, portions of tubes were seen extending in line with the short tubes though apparently not continuous with them, but often with careful focussing a continuity could be traced, though obscured by overlying connective tissue. I am therefore of opinion that the utricular glands in the bitch, as in so many other mammals, lie in the mucosa, some almost vertically, others in various degrees of obliquity, so that, when vertical sections are made, some are cut short across, others longer, whilst others again may be seen in almost their entire length. I conclude that all the glands belong to the type of compound tubular glands, that the apparent differences in length are simply due to the mode in which the glands are cut across in making the section, and that the physiological division proposed by Bischoff into simple mucous crypts and proper tubular glands cannot be supported. (Pl. 1, fig. 6, 7.)

I have also examined the non-gravid uteri of the Badger (*Meles taxus*) and *Paradoxurus pallasii* with the object of seeing whether they did or did not correspond with the bitch in the arrangement of the utricular glands. In the badger, when vertical sections were made through the mucosa, an appearance of short simple tubes and of long compound tubes was recognised without difficulty, but on closer examination it was seen that between the shortest and longest tubes every gradation of

length existed, so that I came to the same conclusion, as with
the bitch, that the short tubes were merely utricular glands cut
across in making the section. In *Paradoxurus* the glands had
a similar arrangement.

Fox.—From a dissection which I have made of the gravid
uterus of a Fox at about the mid-period of gestation, I have satisfied myself that it corresponds in many respects with the bitch,
though with specific differences. A layer of mucosa remained
on the uterus when the placenta was stripped off, and possessed
pits or trenches with intermediate ragged folds. The uterine
face of the placenta was flocculent, owing to the prolongations of
these folds into the substance of the placenta being torn across
in the process of separation. These prolongations entered the
placenta at a number of points, and passed with a sinuous
course up to the chorion, dividing in their course into numbers
of trabeculæ, which formed a meshwork, in the meshes of
which the lateral offshoots of the villi were lodged. The
intra-placental laminæ and trabeculæ were very vascular, and
their vessels, which were larger than ordinary capillaries, formed
an anastomosing network. Compared with the capillaries of
the fœtal villi they were from twice to four times as big, so
that they may be spoken of as colossal capillaries, and undoubtedly represent the early stage of a dilatation into maternal
blood-sinuses, such as is still more clearly seen in the sloth,
and reaches its maximum development in the human placenta.
Owing to the subdivision of the maternal laminæ, the trabeculæ usually contained only a single colossal capillary; and as
many of these vessels ran vertically through the placenta, when
horizontal sections were made through the organ they were
seen in transverse section. In many cases these transversely
divided vessels were surrounded by little more than a ring
of cells—the epithelial investment of the trabecula of maternal
tissue in which the vessel lay—the sub-epithelial connective
tissue of the trabecula being so attenuated as to be scarcely
perceptible, and at times even not visible, so that the epithelial
cells rested upon a very delicate adventitia enveloping the wall
of the colossal capillary. The epithelial cells investing the
intra-placental laminæ and trabeculæ were remarkably large
and distinct, and on the average about ¼th or even ⅓rd as large

as the corresponding cells in the bitch. The fox therefore, like the bitch, has no continuous layer of modified mucosa, such as is seen in the cat, on the uterine face of the separated placenta.

The villi of the chorion were broad laminæ, deeply cleft, so as to assume an arborescent arrangement, and gave off both lateral and terminal offshoots in which a compact network of capillaries ramified. (Pl. I. Figs. 4, 5.)

Seal.—I have studied the structure of the zonary placenta of the *Pinnepedia* in the Grey Seal, *Halichœrus gryphus*, a specimen of which, in the sixth month of gestation, I examined in 1872[1]. (Pl. II. III. Figs. 11—16.)

In this animal the inner or fœtal face of the placenta possessed a convoluted appearance, with intermediate depressions or sulci, and the convolutions ran parallel to each other. When a vertical section was made through the placenta and adjacent part of the uterine wall, and the placenta gently drawn away from the uterus, its uterine face was also seen to be convoluted, with the convolutions and sulci in reverse order to those seen on its chorionic aspect. A well-defined layer of mucous membrane, which, from its position, represented the serotina, intervened between the muscular coat of the uterus and the placenta, and followed closely the windings of the convolutions, dipping down into the primary fissures between the convolutions in the form of broad laminæ, just as the pia mater dips between the convolutions of the cerebrum. Each convolution was split up into elongated plates by secondary fissures, into which processes of the mucosa, derived not only from the broad laminæ just referred to, but from that in contact with the uterine face of the convolutions, penetrated. Each plate was again subdivided by tertiary fissures into small polygonal lobules, into which more delicate processes of the mucosa entered, and these could be traced through the thickness of the placenta up to the chorion.

The mucosa could readily be peeled off the uterine face of the placenta, as in the Dog and Fox, and when this was done the laminæ were drawn out of the primary fissures, just as one can draw the pia mater out of the cerebral sulci when the grey matter

---

[1] *Trans. Roy. Soc. Edinburgh*, 1875.

on the surface of the cerebrum is exposed. The more delicate processes, however, which entered the secondary and tertiary fissures were torn through in the act of peeling, and remained in the substance of the placenta entangled between the fœtal villi. When these processes were seized with a pair of fine forceps, and gentle traction employed, they could be withdrawn without much difficulty from the substance of the placenta. From the ease with which the processes lying in the secondary and tertiary fissures tore across in the act of peeling off the placenta, there could be little doubt that a similar disruption occurs in the separation of the placenta during normal parturition. This opinion was confirmed by an examination of the placenta of a *Phoca vitulina*, shed at the full time, which is in the Museum of the College. The part of the mucosa, therefore, which is shed along with the placenta, consists of the delicate easily torn processes which dip into the secondary and tertiary fissures, and are entangled between the placental lobules and amidst the fœtal villi. The uterine face of the separated placenta is not covered by a continuous layer of decidua, for the greater part of the mucosa is not shed when the placenta is expelled, but remains as a layer of membrane of considerable thickness on the inner surface of the muscular coat, and presents on its placental aspect numerous irregular pits or trenches into which the convolutions of the placenta are received when the organ is *in situ*.

Before describing the minute structure of the serotina, I shall relate some observations on the structure of the mucosa in the non-gravid uteri of some seals, which I have had the opportunity of examining, and also the structure of the uterine mucosa in the gravid uterus of *H. gryphus*, both in the non-gravid horn and in the non-placental area of the gravid horn.

In the non-impregnated uterus of a young seal (species unknown) elongated, tubular, utricular glands were very numerous, and closely packed together in the mucosa. The glands lay perpendicular to the plane of the surface, were tortuous, and apparently branched at their deeper ends; by their opposite extremities they opened by funnel-shaped mouths on the free surface of the mucosa. They were lined by a columnar epithelium, and possessed a central lumen. The interglandular connective tissue contained multitudes of corpuscles.

In the uterus of an adult non-gravid Grey Seal the mucous membrane formed strong folds extending in the longitudinal direction. By its deep surface this membrane was connected to the muscular coat by a lax connective tissue. Vertical sections through the mucosa, examined microscopically, displayed numerous tubular glands, which opened freely on the surface. Their main stems lay almost perpendicular to the plane of the surface, but as the glands were somewhat tortuous, and gave off lateral offshoots, they were not unfrequently transversely or obliquely divided. The epithelium did not fill up the gland-tubes, but left a central lumen. The exact form of the epithelium cells could not definitely be made out, but the end which lay next the lumen was rounded, or somewhat polygonal, like the broad free end of a columnar epithelium cell. The interglandular connective tissue was vascular, and a well-marked capillary plexus ramified immediately beneath the surface of the mucosa around the mouths of the glands.

In the uterus of a *Cystophora cristata*, which died about three and a half months after the birth of a cub, I was able to confirm the observations previously made on the uterus of *H. gryphus*. On the free surface of the mucosa was a layer of columnar epithelium, the cells of which were large and well-formed. The mouths of the utricular glands were seen without difficulty opening on the summits and sides of the longitudinal folds of the mucous membrane; their orifices were circular, closely set together, and each was surrounded by a capillary vascular ring. The free surface of the mucosa was thus studded with multitudes of minute orifices—the mouths of the glands. The glands were comparatively short both in *H. gryphus* and *C. cristata*, and the capillaries of the mucosa formed a closely-set network around them.

The free surface of the non-gravid horn of the pregnant uterus of *H. gryphus* possessed no longitudinal folds of its mucous membrane such as were observed in the non-gravid uteri of *H. gryphus* and *C. cristata*. The surface of its mucosa was to the naked eye almost perfectly smooth, but when examined with a simple lens, slight irregularities were seen, partly due to the presence of minute ridges with intervening depressions, and partly owing to a granulated condition of the membrane. When examined with higher powers of the microscope the tubular utricular glands were readily seen. They were more elongated, and less tortuous than in the unimpregnated uterus; the branches at their deeper ends were much more distinctly seen, and they were much less closely crowded together, owing to the increase in the amount of the interglandular connective tissue. The glandular epithelium was abundant, the cells being elongated, though I could not satisfactorily determine that they possessed a precise columnar form. The granulated appearance of the mucosa seemed to some extent due to the presence of these glands in the membrane.

The free surface of the mucous membrane of the non-placental area of the gravid horn of the same uterus was smooth in appearance, both to the naked eye and under a simple lens. With higher powers the tubular utricular glands were also seen without difficulty, but they were more elongated, so slightly tortuous as in many instances to be almost straight, and separated by greater intervals, occupied by the interglandular connective tissue, than in the non-gravid horn. In some of the glands the columnar form of the cells was distinctly recognised, and the almost circular form of the gland-orifice on the free surface of the mucosa was in many preparations readily seen. The mucous membrane of the septum between the two horns was smooth on the aspect directed both to the gravid and non-gravid horn. The appearance and form of the glands, and the proportion of interglandular connective tissue, was almost alike on both aspects.

From a comparison of the mucosa of the gravid with the non-gravid horn of the pregnant uterus, and of these with the unimpregnated uteri in these seals, it is evident that the changes which take place in the mucous membrane, in connection with the great distension of the uterus during pregnancy, consist in an obliteration of the strong longitudinal folds of the mucosa; in a large increase in the absolute and relative amount of the interglandular part of the mucous membrane; in an elongation of the tubular glands, which elongation is in great part due to an untwisting of the glands, so that they become

T.   7 much less tortuous, though from the very considerable length which some of these glands possessed, it is possible that they and their branches may have actually grown in length. The similarity in the appearance of the septal mucous membrane on its two aspects was evidently due to the growth of this partition being equal for the non-gravid as for the gravid horn.

A broad band of mucous membrane reflected on to the border of the placenta (*decidua reflexa*) was smooth on its free surface, like the adjacent part of the uterine mucosa with which it was continuous. When peeled off the placenta, and placed under the microscope, utricular glands were seen in it, which in form and relative numbers closely corresponded with the arrangement just described in the mucosa of the non-placental area of the gravid horn. Many of the glands, however, displayed an appearance such as I had not previously observed; for their lumen, instead of being empty, was occupied by a bright yellow material, which probably was the secretion confined within the lumen through some obstruction near the mouth of the gland preventing its excretion.

I then proceeded to examine the structure of the mucosa in the placental area. In the non-deciduous serotina, *i.e.* in the layer of mucous membrane left on the wall of the uterus after the placenta was peeled off, utricular glands were seen, but they were much more sparingly distributed even than in the mucosa of the non-placental area of the gravid horn. In various of these glands an appearance was observed, indications of which had also been seen in some of the glands both in the mucosa of the non-placental area and in the reflexa, of a breaking up within the gland-tubes of the epithelium into scattered masses, separated by intermediate irregular intervals.

On that surface of the non-deciduous mucosa, which was exposed by peeling off the placenta, irregular scattered patches of cells were seen when examined with a magnifying power of 300 diameters, which were obviously portions of the originally continuous epithelial layer of the mucosa, that had become broken up into patches by the removal of some of the cells in the act of peeling off the placenta. It was observed that the cells remained in position on those parts of the mucosa immediately superficial to its larger blood-vessels, whilst they were frequently absent from the surface of the membrane situated between these vascular trunks. The cells in a patch were in close contact with each other. They were short columnar cells; their free ends being either circular, or ovoid, or polygonal, and in many cases having the diameter of a white blood-corpuscle, though others were somewhat larger.

Both the non-deciduous serotina and the decidua reflexa were much more vascular than the mucosa of the non-placental area of the uterus. The increased vascularity was due to the blood-vessels being larger, and apparently more numerous in a given area. In all these localities vessels of capillary size were present, but the veins and arteries of the serotina and reflexa were considerably larger than those of the non-placental mucosa. This increase in size was not due to the formation of varicosities on limited portions of their walls, but to a general expansion of the vascular tube. No curling or cork-screw-like arteries were seen, and the veins presented no unusual tortuosity. The sub-epithelial connective tissue contained multitudes of well-marked connective-tissue corpuscles.

The broad laminæ of mucosa which dipped into the primary fissures between the convolutions of the placenta had an interrupted layer of epithelial cells on their free surface, similar in shape but somewhat bigger than those of the non-deciduous mucosa just described. The arrangement and relative size of the blood-vessels were also the same, and utricular glands were present, though sparingly distributed in the sub-epithelial connective tissue.

The structure of the delicate bands of deciduous mucosa which passed into the secondary and tertiary fissures in the substance of the convolutions was then examined. The free surface of these bands was covered by an epithelial layer, the cells of which were columnar like those of the non-deciduous mucosa; but their contents were more opaque and yellow, as if in process of fatty degeneration. Flake-like layers of cells were not unfrequently seen lying loose in the fluid in which these specimens were examined, as if they had become detached from the free surface of the decidua. In one or two instances rows of cells, as if the cellular contents of utricular

glands, were observed, but no glands were seen in these delicate processes. The bands of decidua, lying in the secondary
and tertiary fissures, consisted of a delicate membranous connective tissue, into which the blood-vessels of the non-deciduous
mucosa were prolonged. These decidual bands dipped between
the lobules of the placenta, almost up to the chorion, and the
maternal vessels branched and formed in them a capillary network. From these bands slender processes passed into the
interior of the placental lobules, where they formed a lattice-
like arrangement of very slender trabeculæ, winding in a sinuous
manner through the lobule. These trabecles could be pulled
out of a lobule with a pair of fine forceps, and in many specimens the bud-like processes of the chorionic villi, lodged in the
interstices between the trabecles, were drawn out along with
them. Each trabecle was formed of a capillary blood-vessel,
surrounded by a thin layer of connective tissue, which again
was invested by a layer of columnar epithelial cells similar to
those already described on the free surface of the bands of decidua. These cells were very easily detached from the surface
of the trabeculæ, and quantities of loose cells floated about the
fluid in which the specimens were examined. The deep attached end of a cell was often attenuated into a fine process.
The trabecles were, therefore, delicate bands of the uterine
mucosa, and were composed of its several constituents *minus*
the utricular glands.

When the uterine face of the placenta, from which the non-
deciduous mucosa had been peeled off, was examined, a greyish
membrane was seen, which lay in contact with the uterine face
of the placental lobules. It was not, however, prolonged from
the uterine surface of one lobule to the corresponding surface
of the adjacent lobules, so as to form a continuous layer over
the whole uterine surface of the placenta, but was continued
for some distance down the side of each lobule into the substance of the placenta, so as to form an investment for the
individual lobules. Hence the uterine face of the placenta was
broken up into polygonal areas, each of which corresponded to
a placental lobule, and the areas were separated from each
other by bands of the uterine mucosa, which dipped into the
secondary and tertiary fissures of the placenta. The greyish

membrane belonged to the fœtal and not to the maternal part of the placenta; for whilst the mucosa readily peeled off from the one surface of the grey membrane, the other surface was continuous with the tissue of the villi, and could not be separated without tearing through small blood-vessels and connective tissue which passed from the villi into it.

The greyish membrane was composed of connective tissue, the corpuscles of which were ovoid and fusiform, relatively large in size and granular. Ramifying in the membrane were small blood-vessels continuous with those of the chorionic villi, which had been torn through when the grey membrane was stripped off. On that surface of the greyish membrane which lay next the non-deciduous mucosa, patches of epithelial cells similar to those previously described on the free surface of the mucosa were seen. I believe that these cells, though adhering to the membrane, did not properly belong to it, but to the mucosa, from which they had separated in peeling off the placenta; and in this manner one may explain why the epithelial covering of the mucosa seemed to form an interrupted and not a continuous layer.

Numerous vertical sections were now made through the entire thickness of the placental lobes, and examined with the view of determining the arrangement and structure of the villi of the chorion, their more exact connection with the greyish membrane, and their relations to the intra-lobular parts of the mucosa. The stems of numerous large villi arose at frequent intervals from the placental surface of the chorion, and passed through the placental lobules almost perpendicular to the plane of the chorion, and branched in a highly arborescent manner. From the sides of the stems of the villi, from the sides of their branches, and from the extremities of the greater number of these branches, much smaller branched villous processes arose which gave origin to multitudes of villous tufts. Some of the larger branches from the parent stem had, however, a different mode of termination: they reached the periphery of the lobule and blended with the greyish membrane, which was obviously formed by the junction with each other of the ends of those branches of the villi which reached the periphery of the lobule; and by their union a continuous layer of fœtal tissue was formed, not only on the uterine surface of each lobule, but reaching for some distance down its sides. From the placental surface of the chorion, in the intervals between the origins of the stems of the large arborescent villi, numbers of short branching villi arose, which soon subdivided into terminal branching tufts. The terminal branching tufts were, as a rule, slender elongated structures, but some were shorter and more club-shaped.

The matrix substance of the villi consisted of a delicate connective tissue containing multitudes of distinct corpuscles. Where this tissue formed the terminal tufts the corpuscles were very numerous, and appeared in some cases not only imbedded in the substance of the tuft, but as if arranged, after the manner of an epithelium, on the free surface. In some of my preparations the more superficial cells were detached, and were seen to have the form of delicate scales of nucleated protoplasm. The larger blood-vessels lay in the stems of the villi and branched in an arborescent manner, ending in a capillary plexus in the terminal villous tufts. In some of the larger branches of the villi, a vessel ran parallel and close to the surface of the villus, which communicated with the capillaries of the tufts arising directly from the sides of the branches.

The intra-lobular prolongations of the maternal mucosa did not pass uninterruptedly into the lobules, for the greyish membrane situated on their uterine surface, and on the adjacent part of the sides of the lobule, prevented a direct entrance. The intra-lobular decidua was, therefore, derived from those processes of the mucosa which dipped into the secondary and tertiary fissures. These processes, in the form of slender bands and laminæ, penetrated up to the chorion, and then branched off laterally into the lobules, when they at once broke up into the reticulated lattice-like arrangement of sinuous trabeculæ already described. In sections made through the lobules, where no displacement of the relative position of the fœtal and maternal structures had taken place, the meshes of the reticulum were seen to be occupied by the villous tufts, and not unfrequently the tufts were surrounded by a ring-like arrangement of trabecles. In this manner, throughout the entire lobule, the maternal and fœtal parts of the placenta were so closely intertwined that the two systems of blood-vessels were brought into close juxtaposition with each other: the structures which intervened being, on the maternal side, the epithelial investment of the trabeculæ, and on the fœtal, the flattened scale-like superficial cells of the villi.

From the mode in which the placental lobules were walled in on the uterine aspect by the greyish membrane of fœtal tissue, from the processes of decidua having to penetrate up to the chorion before their capillaries entered the lobules, and from the recurrent course which so large a proportion of the intra-lobular trabeculæ had to take in order to reach the villi situated nearest to the greyish membrane; the maternal blood-vascular system penetrated throughout the entire lobule, and was brought into relation with the numerous offshoots of the villi. In the separation of the placenta during parturition a quantity of maternal vascular tissue comes away therefore with the placenta.

The seal, in the reticulated arrangement of those portions of its mucosa which are in direct contact with the terminal villi, presents a general correspondence with the fox, but the subdivision of the mucosa is more complete, and the cells of the epithelial investment are not so big as in the fox.

Hyrax.—The zonary form of the placenta in *Hyrax capensis* was pointed out many years ago by Sir Everard Home [1], and the presence of the sac of the allantois equal in length to the chorion was figured though not described by him. Although the structure of its placenta has since been examined by several anatomists, there is by no means an agreement on the exact relations of its fœtal and maternal portions. Prof. Owen refers, in his description of the placenta of the Elephant, to the placenta of Hyrax as similar in spongy texture and vascularity to the annular placenta of the *Carnivora*, and subsequently states [2] that the villi are imbedded in a decidual substance, and the surface of attachment to the uterus is less limited than in the Elephant. Prof. Huxley is convinced from his investigations [3]

[1] *Lectures on Comparative Anatomy,* v. 325, and vi. Pl. 61, 62. 1828.
[2] *Comparative Anatomy of Vertebrates,* iii. 742.
[3] *Elements of Comparative Anatomy,* 1864, p. 111. *Manual of the Anatomy of Vertebrated Animals,* 1871, p. 134.

that the placenta in *Hyrax* has such an interblending of the foetal and maternal portions that it is as truly deciduate as that of a Rodent. The maternal vessels, he says, pass straight through the thickness of the placenta towards its foetal surface, on which they anastomose, forming meshes, through which the vessels of the foetus pass towards the uterine surface of the mother. The allantois spreads over the interior of the chorion and gives rise to the broad zone-like placenta. The amnion is not vascular. In the foetus the yelk-sac and the vitello-intestinal duct early disappear. On the other hand, M. H. Milne-Edwards describes[1] the placenta as only adhering very feebly to the walls of the uterus. Its villi, he says, are mostly simple, very analogous to those of an ordinary pachyderm. In the midst of the zone there are vascular vegetations engaged in corresponding uterine cavities, but they adhere no more, than do the analogous prolongations in the Ruminant, to the crypts in which they are included : they can be detached with the same facility without tearing through anything and without carrying away any portion of uterine tissue. There is nothing, he concludes, to indicate the presence of a caduca, and the allantois does not overstep the limits of the placental zone. In an elaborate monograph on the genus *Hyrax* published a few months ago[2] M. George figures not only the placenta but the gravid uterus of this animal. He says nothing however of the structure, which he was apparently precluded from examining, but adopts the view of M. Milne-Edwards that it was nondeciduate.

Owing to the great discrepancy, not only as regards the structural details, but the conclusions as to the nature of the placentation, it was obviously advisable that the placenta of the animal should be re-examined, and with great liberality Prof. Huxley has placed at my disposal his specimen, which had been preserved in spirit. I shall now describe what I have seen[3].

The uterus was two-horned. In one cornu were two well-developed ova, each about $3\frac{1}{2}$ inches long. One ovum had

---

[1] *Considérations sur la Classification des Mammifères*, Paris, 1868.
[2] *Annales des Sciences Naturelles*, June, 1875.
[3] See also *Proc. Roy. Soc. London*, Dec. 16th, 1875.

been opened, the membranes and placenta examined and the foetus removed. The other ovum was entire. The placenta was zonary and varied in different parts of the zone from between ¼ to ½ inch in breadth, whilst the average thickness was $\frac{1}{10}$ inch. The non-placental areas of the chorion were smooth and translucent, and, as in the *Carnivora*, branches of the umbilical vessels ramified in them. The zone on the chorion was intimately united to a corresponding zone in the uterine mucosa. When the placenta was stripped off the uterus, not only was the mucosa in the placental zone removed along with the chorion, but, for some distance on each side of the zone, the mucosa tore away from the muscular coat of the uterus.

The foetal surface of the placenta was smooth, and on it the umbilical vessels ramified before entering the villi of the chorion: some injection was passed into them, but it did not flow into their intra-villous branches. The uterine surface of the artificially separated placenta was flocculent, and the flocculi were seen on microscopic examination to consist of bundles of connective tissue intermingled with corpuscles—obviously the sub-mucous connective tissue of the placental zone.

A toughish membrane, continuous on each side with the mucosa of the non-placental area of the uterus, could without difficulty be peeled off the uterine surface of the placenta as a well-defined layer. When examined microscopically, the surface of this membrane which lay next the uterus was seen to have the structure of the flocculi which projected from it; whilst the part next the placenta was much more abundantly cellular, and stained readily with carmine. The surface which had been in apposition with the placenta, was irregularly undulating, and divided into numerous shallow crypt-like recesses, separated from each other by raised folds of the membrane. These crypts were lined by a layer of cells, the nuclei of which were very distinct. I regard this membrane as the serotina, or modified uterine mucosa in the placental zone, whilst the crypts with their epithelial lining are the closed ends of the deep pits in which the villi of the chorion are lodged.

The substance of the placenta was then examined with the

object of determining not only the form of the fœtal villi, but
if prolongations of the maternal mucosa passed into the placenta between the villi. In vertical sections the villi appeared
as if simple filamentous structures, extending from the chorion
to the crypt-like recesses in the serotina. But when portions
of the placenta were teased out with needles, and the villi not
injured, although an occasional simple villus was seen, the
majority were broad, sinuous, leaf-like villi, with bud-like
offshoots at the free borders, such as I have described and
figured in the cat (Pl. I. fig. 3). In horizontal sections
through the placenta, the sinuous outline of the villi had so
close a resemblance to the cat, that it was difficult in this particular to distinguish *Hyrax* from *Felis*. Under a magnifying
power of 320 diameters, a layer of flattened cells, the nuclei in
which were almost circular and bright with transmitted light,
was seen at the surface of the villus. Intermingled with the
fœtal villi were laminar prolongations of the maternal mucosa,
which invested and closely followed the sinuous outline of the
villi, similar to the arrangement described in the domestic cat
(p. 76 and fig. 2). This intra-placental maternal tissue was present not only in those sections in which the serotina was *in situ*,
but in others from which this membrane had been peeled off.
The maternal laminæ were covered by a layer of cells, obviously
continuous with the cellular lining of the crypt-like recesses on
the free surface of the serotina, and the laminæ themselves
were prolongations into the placenta of the folds of the mucosa
which separated the crypts from each other.

As the placenta of *Hyrax*, in both the form of its villi and
the mode in which they are interlocked between the intra-
placental maternal laminæ, so closely resembles the domestic
cat, and as these laminæ remain *in situ*, after the membrane,
which I have named the serotina, is peeled off the placenta,
there can be no doubt that they are shed at the time of separation of the placenta. Hence *Hyrax* in its placentation is one
of the Deciduata. Whether the membrane just referred to is
also shed during parturition is more difficult to say. The fact
that it peels off the uterus along with the placenta, when they
are artificially separated, is not of itself sufficient evidence, and
it may be that in *Hyrax*, as in the Cat, the superficial portion

only of this membrane falls off with the placenta, whilst the rest remains on the zone of the uterus. The question, however, can only be decided by the examination of a uterus immediately after parturition.

The non-placental area of the mucosa contained tubular glands, which relatively to the extent of the membrane, were not nearly so numerous as in the gravid mucosa of animals having a diffused or polycotyledonary placenta. The glands were long, and the gland-stem was unbranched and almost straight: the closed end was dilated, bent on itself, and seemed to give off a short branch. The secreting epithelium did not fill up the tube, but left a central lumen.

When the chorion was cut through on either side of the placental zone, the sac of the allantois was opened into and seen to extend up to the poles of the chorion. At and close to the margin of the placenta the allantois was reflected on to the outer surface of the amnion, to which it obviously gave a complete investment, so that the bag of the amnion was suspended in the sac of the allantois by the bands of the latter membrane reflected on to its outer surface. The amnion was large relatively to the sac of the allantois. The umbilical cord was short and somewhat flattened.

Numerous flattened plates, the largest of which were about $\frac{1}{16}$th inch in diameter, projected from the inner surface of the amnion. They were situated, not only in the neighbourhood of the umbilical cord, but scattered as far as the poles. Some were sessile, others were slightly pedunculated, and as they were all covered by the epithelial lining of the amnion, they were apparently developed between it and the layer of allantois which covered its outer surface. The plates were obviously composed of cells, the nuclei of which were distinct, but the toughness of the tissue, due probably to the prolonged action of the spirit in which the specimen had been preserved, rendered it difficult to isolate them. In position and structure these plates apparently corresponded to the amniotic corpuscles which I have described (p. 23) in *Orca gladiator*.

The mucous membrane of the non-gravid uterine cornu was greatly hypertrophied in sympathy with the changes in the gravid horn. There was no extension of the foetal mem-

branes into this cornu, and the cavity seemed as if almost
obliterated by the thickening of the mucous membrane. *Hyrax*
agrees, therefore, with *Felis* not only in the form and structure
of the placenta, but in the large size of the sac of the allantois;
it differs in the condition of the umbilical vesicle, which disappears in *Hyrax* apparently at an early period, but remains in
Felis to the end of utero-gestation.

Elephant.—No observations have been recorded on the
structure of the uterine mucosa in the gravid Elephant, but
Prof. Owen has described and figured[1] the fœtal membranes
at about the mid-period of utero-gestation. The chorion was
encompassed at its middle by an annular placenta, 2 ft. 6 in. in
circumference, varying from 3 to 5 in. in breadth, and from
1 to 2 in. in thickness:

"The placenta presents the same spongy texture and vascularity
as does the annular placenta of the *Hyrax* and of the *Carnivora;* but
the capillary filaments or villosities enclosing the fœtal vessels enter
into its formation in a larger proportion, and are of a relatively
coarser character. The greater part of the outer convex surface of
the placenta is smooth; the rough surface, which had been torn from
the maternal or uterine placenta, exposed the fœtal capillaries,
and occupied chiefly a narrow tract near the middle line of the
outer surface. A thin brown deciduous layer is continued from the
borders of the placenta, for a distance varying from one to three
inches, upon the outer surface of the chorion. In addition, at each
of the poles of the chorion was a villous and vascular subcircular
patch, between two and three inches in diameter, the villi being short,
⅙th of a line in diameter, or less.

"The bag formed by the mucous or unvascular layer of the allantois is of considerable size, is continued from the base of the umbilical
cord, so expanding between the chorion and amnios as to prevent any
part of the amnios attaining the inner surface of the placenta. The
allantois divides, where the amnios begins to be reflected upon it, into
three sacculi; one extends over the inner surface of the annular
placenta and a little way into one end of the chorion: a second
extends into the opposite end of the chorion, it there bends round
toward the placenta, and its apex adheres at that part to the first
division of the allantois: the third prolongation subdivides into two
smaller cavities, each terminating in a cul-de-sac, encompassing, and
closely attached to the primary divisions of the umbilical vessels."

This specimen is preserved in the Museum of your College,
and through the courtesy of Professor Flower I have been

---

[1] *Phil. Trans.* 1857, p. 347, and *Comp. Anat. of Vert.* III. 740.

permitted to obtain a slice of the zone for microscopic examination. The size and thickness of the zone prove it to be the functionally active part of the placenta, for the villous patch at each pole is so small as to be of little physiological importance. Notwithstanding the number of years the placenta had been in spirit, I succeeded in passing some injection into the vessels of the chorion and the larger trunks in the stems of the villi, so that I could follow the villi more precisely into the substance of the placenta than I should otherwise have been able to do. The placenta was very compact and was clearly composed both of a fœtal and a maternal portion closely interlaced with each other. Many of the villi were of large size and passed through the entire thickness of the organ, branching repeatedly in an arborescent manner. Others again were of smaller size, and did not pass more than one-third through the organ, but, like the longer villi, branched repeatedly. The tissue of the villi was delicately fibrillated, and in it ran the branches of the umbilical vessels.

Interlocked between the villi was a tissue, which contained a very distinct network of minute tubes, obviously capillary blood-vessels, and on the surface of this tissue a layer of cells was seen with some difficulty. I succeeded more than once in isolating a few of these cells, and found them to be rounded or ovoid, with definite nuclei and with granulated protoplasm. I believe these cells to be the epithelial covering of the laminæ of maternal mucosa, forming the walls of the highly-developed crypts in which the villi were lodged, whilst the capillary network subjacent to these cells belonged to the intra-placental maternal vascular system. Several times I saw an appearance as if the intra-placental mucosa was split up into a reticulated arrangement of trabeculæ, similar to what I have described in the seal, but from the condition of the specimen it was difficult to speak positively on this point. There could be no doubt however that in this separated placenta of the elephant a large amount of uterine mucosa was inextricably locked in between the fœtal villi.

GENERAL MORPHOLOGY OF THE DIFFUSED, THE POLYCOTY-
LEDONARY, AND THE ZONARY PLACENTA.

I shall now proceed to make some general observations on
the Morphology of the Placenta. In the study of the morphology of the placenta in any mammal the presence of two
parts, a fœtal and a maternal, originally quite distinct and
separable from each other, must be clearly kept in view.

The morphology of the fœtal part presents no difficulty. It
consists simply of a vascular villous membrane covered by an
epithelium. The sub-epithelial part of the membrane is composed of a delicate connective tissue, containing numerous
corpuscles, in which the terminal branches of the umbilical
vessels, with their capillary network, are distributed. The
vascular villi may be either simple or branched, and in some of
the mammals, whose placentation has just been described, e.g.
the seal, the branching may assume a highly arborescent
arrangement.

The morphology of the maternal part of the placenta presents greater difficulty, not only because the uterine mucous
membrane, out of which it is produced, is more complex in
structure than the chorion, but because this membrane becomes greatly modified in the course of placental development,
and not unfrequently becomes so interlocked between the
fœtal villi as to be separated from them with great difficulty.

In all the forms of placenta which have just been described,
along with the growth of the villi from the surface of the
chorion, depressions or Crypts arise in the uterine mucosa for
their reception, and the walls of these crypts are formed by
foldings of the hypertrophied mucosa.

In the Diffused placenta the changes in the uterine mucosa
are less complicated than in the other forms. The villi of the
chorion are short, and branch but slightly. The crypts in the
uterine mucosa are consequently shallow, so that the relations
of the fœtal and maternal parts can be easily seen. Two free

surfaces are in close apposition, the villi of the chorion fit into the crypts of the mucosa, but they can be drawn asunder without difficulty, as the hand is drawn out of a glove; so that the compound nature of the placenta can be at once demonstrated.

In the Polycotyledonary placenta the villi are longer and more branched. The pits or crypts for their reception are consequently deeper and divided into smaller compartments, and the maternal mucosa in the site of the cotyledons is more hypertrophied, thicker, and more spongy. Two free surfaces are here also in apposition; but the length and branching of the villi, and the depth and subdivision of the crypts, render it somewhat more difficult to draw the two surfaces asunder than in the diffused placenta.

In the Zonary placenta as seen in the *Carnivora*, *Pinnepedia* and *Elephas*, the villi are long and usually arborescent, though in the Cat and *Hyrax* they are leaf-like and very sinuous. The foldings of the uterine mucosa, which have led to the production of the crypts, are more complicated, so much so indeed in the Fox and Seal, as to give rise to a remarkable subdivision of the membrane into a microscopic network. The two surfaces in apposition have become so interlocked that it is almost impossible to disengage them from each other. Hence in the process of parturition more or less of the uterine mucosa in the placental area is separated and shed in the substance of the placenta.

The morphological elements in the gravid mucosa of all mammals, are, as in the non-gravid membrane, epithelium, sub-epithelial connective tissue, blood- and lymph-vessels, glands and nerves. Of the arrangement of the lymph-vessels and nerves in the placenta we have no precise information. The epithelium, the sub-epithelial connective tissue, and the blood-vessels form the walls of the crypts in which the villi are lodged. The glands have no necessary relation to the crypts. In the Pig, as has been shown by Eschricht, myself, and Ercolani; in the Mare, as has been pointed out by Ercolani and myself; in the Porpoise, as has been described by Eschricht; in the Narwhal, as I have occasionally seen; in the Lemurs, as I have also described,—the mouths of the glands can be distinctly seen

opening on the surface of the mucosa, in smooth areas surrounded by and quite distinct from the crypts. In *Orca
gladiator*, though at first sight the funnel-shaped crypts seemed
to be the dilated mouths of glands, further consideration
has satisfied me that neither they nor the cup-shaped crypts
are derived from the glands. In the *Ruminantia*, Eschricht,
Bischoff, Ercolani and I have been unable to see any communication between the glands and the pit-like crypts of the
cotyledons. In the *Carnivora*, though, as was interpreted by
Sharpey, Weber, and Bischoff, the crypts seen in the placental
area in the early stage of gestation may seem to be merely
the mouths of the glands enlarged and widened, yet a more
minute analysis of the structure shows that, though some of
the glands may, as in *Orca*, open into crypts, the crypts
are much more numerous than the glands, and are consequently not derived from them. In the *Pinnepedia*, *Hyrax*
and the Elephant, there is also no reason to believe that
the crypts are formed by a widening of the glands. Hence
in all these, and I believe in other placental mammals, the
crypts are not modified glands, but are inter-glandular in position. The crypts do not exist in the non-gravid uterus, but,
as was first definitely shown by Prof. Ercolani, from the examination of the placenta, in several orders of mammals, are formed
during pregnancy by foldings of the mucous membrane.

The crypts are lined by an epithelium, which is derived
by descent from, and lies in the same morphological plane as,
the epithelial lining of the uterus: the increase in the number
of epithelial cells, required by the much greater magnitude of
the mucous surface, being effected by proliferation of the pre-
existing epithelial cells. In many mammals the cells lining
the crypts have the columnar form, like the epithelium of the
non-gravid mucosa, and in the pig the cells are apparently
ciliated: but in some mammals the columnar form is not preserved, and the cells are rounded, or polygonal, and with granulated protoplasm. These cells form the cells of the decidua
serotina, and they are homologous with the rounded, or polygonal, colossal, granulated cells of the decidua serotina in the
human placenta.

The connective tissue in the walls of the crypts is derived

from the sub-epithelial connective tissue of the non-gravid mucosa, through a rapid increase in the number of its corpuscles, though it is possible that there may also be a migration of white blood-corpuscles into it. The blood-vessels in the walls are continuous with the vessels of the mucosa, and are greatly increased in numbers. In the diffused and polycotyledonary forms of the placenta they are arranged as a capillary network, but in the zonary placenta they exhibit a tendency to dilate into colossal capillaries, which are the first indications of a maternal intra-placental blood-vascular sinus system, such as attains much greater development in the Sloth, and acquires its maximum size in the *Quadrumana*, and the Human Female. The vascular connective tissue forming the walls of the crypts constitutes the vascular part of the Decidua Serotina, by which term is signified the maternal mucous membrane situated between the fœtal placenta and the muscular wall of the uterus; or, in other words, the maternal part of the placenta. In the diffused form of placenta the Serotina consists of the whole of the mucous surface in which the crypts are met with. In the polycotyledonary it forms the maternal cotyledons. In the zonary placenta it consists of the annular band of mucosa, with the intra-placental laminæ and trabeculæ.

As is well known, the form of the placenta, the arrangement of the fœtal membranes, and the behaviour of the uterine mucosa at the time of parturition, have been taken by many zoologists as affording a basis of Classification of the placental mammals.

Fabricius was well aware[1] that the placenta varied much in shape, size and position, in different mammals, and that a particular form of placenta was also associated with certain other anatomical characters. Thus, he says, the single placenta found in the human female, the mouse, rabbit, guinea pig, dog and cat, is associated with the presence of incisor teeth in both jaws and with distinct toes. In some of these mammals, as the human female, rabbit, hare, mole, mouse and guinea pig, the single placenta resembles a cake (whence indeed its name placenta); in others, as the dog, cat and ferret, the

---

[1] *De Formato Fœtu.* Folio ed. 1624. Part 1, Caput III. p. 4.

placenta is like a girdle or zone. When the placenta is multiple, as in the sheep, cow, goat or deer, the incisors are present in one jaw only, and the hoofs are cloven. These observations by Fabricius are quoted, with approbation, by the illustrious Harvey[1], and they express very clearly the conjunction of certain well-defined characters with a particular form of placenta. In 1823 Sir Everard Home proposed[2] an arrangement of the Animal Kingdom founded on modifications of the egg, and laid down certain characters of the mammalian orders taken from the structure of the placenta. Although some, which he gives, are sufficiently precise, as for example the belt-like form of the placenta in the *Carnivora*, and the cotyledonary form in the *Ruminantia*, yet in other respects the characters he has laid down are too indefinite to be of much value.

It was not until a few years later, when von Baer published the first part of his most important treatise on the Development of Animals, that a definite system of Classification based on their development was propounded. In his *Entwicklungsgeschichte*[3] he divides the placental mammals into groups, according to the size of the umbilical vesicle and allantois, as in the following Table:

The Umbilical Vesicle	persists. The Allantois grows	very little. *Rodentia.*
		moderately. *Insectivora.*
		strongly. *Carnivora.*
	grows little. The Allantois	grows little. Funis very long. *Apes, Man.*
		persists. Placenta — in isolated clumps, or cotyledons. *Ruminantia.*
		diffused. *Pachydermata, Cetacea.*

In his *Untersuchungen*[4] published in the same year, 1828, he added an important character to this classification, by

---

[1] Works, translated by Dr Willis, p. 564.
[2] *Lectures on Comparative Anatomy*, III. pp. 470 and 501.
[3] *Ueber Entwickelungsgeschichte der Thiere*, Erster Theil, p. 225, Königsberg, 1828.
[4] *Untersuchungen ueber die Gefässverbindung zwischen Mutter und Frucht*, p. 26. 1828.

pointing out that in the diffused and cotyledonary forms the
fœtal placenta was merely applied to the maternal placenta,
whilst in the *Carnivora*, with their zonary placenta, and in
the *Rodentia*, *Insectivora* and *Man*, where the placenta is at
one end of the ovum, the fœtal and maternal elements were
grown together. Von Baer had therefore, as Professor Huxley
has already stated, recognised not only important differences
in the form of the placenta in different mammals, but in the
degree in which the fœtal and maternal parts of the organ
were incorporated with each other.

In 1835 Prof. Weber communicated to a meeting of German
Naturalists[1] a classification of the placental mammals in two
groups based on the presence or absence of maternal parts in
the separated placenta: 1st, where the vascular folds or "cells"
of the uterus are so closely attached to the vascular folds or
villi of the chorion, that they are torn through at the time of
birth, so that the uterus is wounded; and the organs which
serve to connect the mother and fœtus fall away at the birth of
the placenta, and they are, as he says, "*hinfällig, organa caduca*,"
e. g. in the bitch, cat, rabbit and man : 2nd, where the uterine
and fœtal parts are so loosely attached that they separate at
birth, without the uterus being torn, just as a sword is drawn
from its sheath, there being no "*zufällige organe*," *e. g.* in the
cow, roe-deer, sheep, stag, mare and pig. Eschricht in his
essay published in 1837[2] employs a similar classification, and
divides placental mammals into two families, in one of which
the uterine placenta is caducous, in the other non-caducous.
M. H. Milne-Edwards published in 1844[3] a system of classification in which he attached great weight to the size and disposition of the allantois and the form of the placenta. In a
subsequent memoir published in 1868[4] he lays stress upon the
presence of a *caduca uterina* in mammals with a zonary or
discoid placenta, and as these animals lose blood at the time
of birth he groups them together under the common term
Hématogénètes. In 1863 Prof. Huxley, in his Lectures on

[1] *Froriep's Notizen*, Oct. 1835, p. 90.
[2] *De Organis*, p. 30.
[3] *Ann. des Sciences Naturelles*, 1844, p. 92.
[4] *Considérations sur la Classification des Mammifères*, Paris, 1868, p. 22.

Classification[1], delivered as Hunterian Professor, suggested
that the terms deciduate and non-deciduate were to be
preferred to caducous and non-caducous, and arranged the
placental mammals into the groups Deciduata and Non-deciduata; an arrangement which has been adopted by several
subsequent writers. By the term Deciduata is meant those
mammals which shed, along with the fœtal placenta, more or
less of the vascular constituents of the maternal mucosa in the
placental area, whilst the Non-deciduata do not part with any
of the mucosa in the act of parturition.

A sharp line of demarcation is therefore drawn by these
eminent anatomists between the mammals whose uterine
mucosa during parturition is caducous or deciduate, and those
in which it is non-caducous or non-deciduate. In employing
these terms it should be distinctly kept in mind that the
same anatomical elements exist in both types of placenta,
and that the shedding or non-shedding of maternal tissue is
determined by the degree of interlacement of the fœtal and
maternal parts of the organ, and not from the presence in
the deciduata of structures which do not exist in the non-
deciduata.

I shall reserve until the concluding course of Lectures the
consideration of the question how far the modifications in the
form and structure of the placenta may be taken as affording
a reliable basis for the Classification of the placental mammals.
On this occasion I shall confine myself to making a few observations on the deciduate or non-deciduate character of the
placentæ described in the present course.

All anatomists agree in regarding the Diffused Placenta as
non-deciduate, for the uterine crypts are so shallow that the
chorionic villi can be drawn out of them with great ease; and
the fœtal membranes are shed in the act of parturition, without
entangling and drawing away the maternal mucosa.

The Polycotyledonary Placenta is also regarded as non-
deciduate. But from observations made on the shed membranes of the sheep and cow, I recently ascertained[2] that intermingled with the villi of the fœtal cotyledons were quantities

---

[1] *Elements of Comparative Anatomy*, London, 1864, p. 103.
[2] *Proc. Roy. Soc. Edinburgh*, May, 1875.

of cells, which possessed the characters of the epithelial cells of the pits and crypts of the maternal cotyledons.

For the purpose of studying the shed placenta of the Sheep, I procured the after-birth from the ewe as soon as it was passed, and immersed it in strong spirit. Some fœtal tufts were then examined without any other preparation; but others were immersed in glycerine jelly, so as to bind the several constituents of the tuft together. Thin slices were then removed from the hardened tufts, whilst from others small portions were taken and teased out with needles. In the examination, a magnifying power of 320 diameters was employed. Quantities of cells, having the form and appearance of the epithelial cells lining the pits in the cotyledons (p. 60), were seen to be intermingled with the fœtal villi. In some cases small patches of cells were seen lying free in the spaces between the villi, but more frequently the cells were isolated. In a few instances I saw groups of such cells in immediate contact with the terminal villi, as if they, in being drawn out of the pits in the maternal cotyledon, had pulled an envelope of epithelial cells along with them.

Fig. 12.

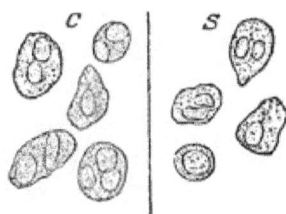

Maternal epithelial cells found intermingled with the fœtal cotyledons in the shed membranes of S the sheep, and C the cow. × 320.

When the cotyledons of the shed placenta of the Cow were examined microscopically, quantities of granular débris were to be seen floating in the fluid in which the specimens were placed. Along with these granules were small flakes of protoplasm; rounded or ovoid bodies, with distinct outlines looking like free nuclei; and large cells composed of granular protoplasm, containing one, two, or three nuclei, having the anatomical characters of the cells lining the pits in the cotyledons. The amount of débris and of decidual cells varies considerably in the different slides which I examined; in some being so abundant as to render the fluid in which the specimen was examined quite turbid, whilst in others only slight traces were to be recognised.

From these observations I am of opinion that, both in the sheep and cow, the cotyledons of the fœtal placenta carry away with them, during the act of parturition, a portion of the maternal structure, so that in these animals, and presumably in other ruminants, the placenta is deciduate. So far as my observations have gone, I have only detected the epithelial element of the uterine mucosa intermingled with the fœtal villi; but from the bloody state of the external parts of the ewe for some hours after the birth of the lamb, I think it not improbable that the disruption of some of the maternal cotyledons has been deeper than a mere epithelial shedding,—that the maternal vessels have, in some places at least, been torn across, so as to give rise to the hæmorrhage.

There is no difference of opinion as to the deciduate nature of the Zonary Placenta in the *Carnivora, Pinnepedia* and Elephant, and although doubts have been thrown by M. H. Milne-Edwards on the deciduate character of the placenta in *Hyrax,* the observations which I have made satisfactorily show that it is undoubtedly deciduate. But it has not been sufficiently recognised that considerable variations occur in the relative proportion of maternal tissue shed along with the fœtal placenta. In the seal, the dog and the fox, the decidua serotina, or mucous membrane of the placental zone, does not form a continuous layer on the uterine face of the separated placenta. A definite layer is however left, when the placenta is shed, on the uterine zone itself, which layer is divided on its surface into pits or trenches by projecting folds. When the organ is *in situ* these folds dip into the substance of the placenta, but are torn through in the process of parturition, so that the only portions of maternal tissue which are shed are the intra-placental prolongations. That the membrane left on the uterus in the placental zone is the mucosa is proved by its vascularity, the layer of columnar epithelium on its free surface, and the utricular glands; which structures, the glands excepted, are also in the intra-placental prolongations. In the feline *Carnivora*, again, as illustrated by the common cat, the mucosa not only sends prolongations into the substance of the shed placenta, but forms a continuous layer on its uterine surface; whilst the layer of tissue left on the placental area of the uterus itself consists not of the entire thickness of the mucosa, but only of the deeper part of the sub-epithelial connective tissue with the remains of glands and blood-vessels. Hence though all the *Carnivora* part with a considerable portion of the maternal mucosa in the separation of the placenta, yet they exhibit differences as regards the degree in which the shedding takes place. The *Felidæ* have a higher grade of deciduation than the *Canidæ*, and with the latter the *Phocidæ* correspond. Hence the Dogs and Seals, in their placental affinities, are less removed from the *Cetacea*, the *Suidæ* and the *Solipedia* than are the Cats. The pits and trenches of the mucosa, which one sees on the uterine zone, after the separation of the placenta in a seal, a fox, or a dog, are obviously similar in their morphological characters to the crypts of the mucosa of a mare, a cetacean, or other animals with a diffused placenta. In the seal the pits and trenches possess a precision of form more than is seen in the dog and fox, a circumstance which is undoubtedly due to the sub-division of the placenta of the seal into definite minute lobules. The higher grade of deciduation in a cat may perhaps be accounted for by the broadly laminated villi, their very sinuous form, and the depth in the mucosa to which their terminal bud-like offshoots penetrate, giving to the fœtal part of the placenta such a "grip," if I may so term it, over the maternal part, as to interlock the latter more firmly with the villi, and thus to cause the mucosa to be to a greater extent shed in the process of parturition.

In the fox and seal the intra-placental prolongations of the mucosa are subdivided into a reticulated arrangement of slender trabeculæ, each bar of which contains only a single dilated capillary; but in the seal this subdivision is carried out to a greater extent than in the fox. In the seal occurs that very remarkable anastomosis of the distal ends of the primary branches of the chorionic villi, which gives to the placenta its precise lobular subdivision, and walls in each lobule at its uterine periphery with the greyish membrane. From a somewhat cursory examination which I have made of the placenta of a *Phoca vitulina*, in the Museum of your College, it appeared to me that a similar membrane existed also in this animal; so that I am disposed to consider the arrangement as one which is of more than generic, indeed of ordinal value.

From the general correspondence in shape and structure between the placenta of the *Pinnepedia* and that of the true *Carnivora*, there can be no doubt that, in both orders, the early stage of formation is marked by the production of crypts in the placental area of the uterine mucosa. In the grey seal the villi of the chorion, which are lodged in these crypts, acquire, not only a considerable length, but a highly arborescent form, and give origin to multitudes of villous tufts. As the branching and growth of the villi proceed in the course of development, the crypts will necessarily become divided into smaller compartments; and as the villous tufts increase in number and size, the walls of the crypts will become no doubt thinned, until at length they will lose their uniformly continuous surface, and become subdivided into the reticulated arrangement already described, in the meshes of the network of which the villous tufts are lodged. That the increased area of the uterine mucosa in the pregnant seal is due to a great increase in the inter-glandular part of the membrane, is proved by the much wider separation of the glands seen in both the non-placental and placental areas of the gravid as compared with the non-gravid uterus of *H. Gryphus*.

It has been customary to regard a placenta as deciduate only when the *vascular* constituents of the uterine mucosa are shed with the foetal membranes. This acceptation of the term seems to me, however, to be too limited, and does not cover all the cases in which maternal tissue is shed in the separated placenta. I would suggest therefore that the definition should be enlarged so as to embrace those cases in which epithelium alone is parted with, as well as those in which both the epithelium and the sub-epithelial vascular uterine tissue come away in the separated placenta. In studying the types of

placenta which have formed the subject of these Lectures we
have passed by successive gradations from the diffused placenta, which is apparently non-deciduate, to the polycotyledonary placenta, in which the epithelial layer of the mucosa only
has been found; then to the zonary placenta of the *Canidæ* and
Phocidæ, where the entire constituents of the intra-placental
prolongations of the mucosa are shed, but where a well-marked
layer of mucous membrane is left on the uterine zone; and
lastly to the *Felidæ*, where, together with the intra-placental
laminæ, the superficial layer of the mucosa in the uterine zone
is shed as a part of the placenta. It follows, therefore, that
the line of demarcation between a diffused non-deciduate, and
a zonary deciduate placenta, is not so sharp as has usually
been supposed, but is graded over by the ruminant polycotyledonary placenta, in which the epithelial layer is the
preponderating if not the only element of the mucosa which
deciduates during parturition. But to prevent misconception
it should be stated, as indeed has been already done by Owen[1],
Ercolani[2], and myself, that if not during parturition, at least
afterwards, all placental mammals are deciduate, for in the pig,
mare, and cetacean, "during the period of involution which follows parturition, it is obvious that great changes, either from
actual shedding of portions of its substance, or from degeneration and interstitial absorption, must take place in the constituents of the crypt-layer before it can be restored to its
proper non-gravid condition[3]." In the Ruminants also, the
thick, vascular, spongy tissue of the maternal cotyledons must
disappear before the uterus can assume its normal uuimpreg-
nated aspect.

[1] *The Anatomy of Vertebrates*, Vol. III. p. 727, 1868.
[2] *Sur les Glandes utriculaires de l'Uterus*, &c.   Algiers, 1869.
[3] *Trans. Roy. Soc. Edinburgh*, 1871, and *Proceedings*, May, 1875.

Physiological Remarks.

THE fœtal placenta possesses an absorbing surface; the
maternal placenta a secreting surface. The fœtus is a para-
site, which is nourished by the juices of the mother.

The absorbing surface of the fœtal placenta is the chorion,
the vessels it contains are the structures which transmit the
materials absorbed to the fœtus, and the villi are the chief,
though not the exclusive, structures engaged in absorption.
In the Diffused Placenta not only the villi, but the smooth
inter-villous part of the chorion, are undoubtedly absorbing
surfaces, for in both a compact capillary network is diffused
beneath the free surface of the membrane. In the Polycotyledonary Placenta, both the fœtal cotyledons and the smooth
inter-cotyledonary part of the chorion are absorbing surfaces, for
not only are the villi highly vascular, but, as I have described
in the sheep, cow, and giraffe, a remarkably compact capillary
plexus is diffused beneath the smooth part of the chorion between
the cotyledons. In the Zonary Placenta the vascularity of the
chorion is not confined to the villi projecting from the annular
equatorial band, but, as I have seen in the cat, bitch, fox,
and grey seal[1], the branches of the umbilical vessels, which are
distributed to the poles of the chorion, terminate in a compact
capillary plexus, and I have little doubt that if the umbilical
vessels in other animals with a zonary placenta were minutely
injected, a similar vascular plexus would be found in them.
Though the villi in the cotyledonary and the zonary placenta
are much fewer in relation to the extent of the chorion than in
the diffused placenta, they are longer, more branched, or more
sinuous, so that the surface for absorption provided by them is
probably as great.

The secreting surface of the placenta is the remarkably
modified mucous lining of the uterus. That the maternal
placenta is a secreting organ, the secretion of which is em-

[1] I have omitted to state, in the description of the placenta of the Grey
Seal, that the umbilical vessels, distributed to the smooth polar areas of the
chorion, terminated in a capillary plexus, as well marked as that which I have
described in the Cat and Bitch.

ployed in the nutrition of the fœtus, has from time to time been definitely stated by various physiologists. The immortal Harvey distinctly recognised that the placenta prepared for the fœtus alimentary matters derived from the mother. In the deer, he says, the maternal cotyledons are

"of a spongy character, and constituted, like a honey-comb, of innumerable shallow pits filled with a muco-albuminous fluid (a circumstance already observed by Galen), and that from this source the ramifications of the umbilical vessels absorbed the nutriment and carried it to the fœtus; just as, in animals after their birth, the extremities of the mesenteric vessels are spread over the coats of the intestines, and thence take up chyle."

And again :

"In my opinion the placenta and carunculæ have an office analogous to that of the liver and mamma. The liver elaborates for the nourishment of the body the chyle previously taken up from the intestines : the placenta in like manner prepares for the fœtus alimentary matters which have come from the mother. The mammæ also, which are of a glandular structure, swell with milk, and although in some animals they are not even visible at other times they become full and tumid at the period of pregnancy ; so too, the placenta, a loose and fungus-like body, abounds in an albuminous fluid, and is only to be found at the period of pregnancy. The liver, I say then, is the nutrient organ of the body in which it is found ; the mamma is the same of the infant, and the placenta of the embryo[1]."

Needham also[2] wrote of the nutritious juice formed in the cotyledons of the *Ruminantia*. Wharton and Haller applied[3] to this fluid the name of milk or milky humor ; by several subsequent writers it has been called uterine milk, and the cotyledons themselves have been regarded as uterine mammæ.

By what structures in the maternal placenta can this fluid be secreted, is a question which must now be considered. E. H. Weber pointed out[4] not only the formation of a chylous fluid from the capillaries of the "cells" (crypts), within the cotyledons of the cow and roe-deer ; but that the secreting surface of the uterus was much increased through the presence of the utricular glands, which opened on the free surface of the mucous membrane between the cotyledons. The secretion of

[1] *The Works of Harvey*, translated by Dr Willis, pp. 562, 563.
[2] *Disquisitio Anatomica*, 1667, p. 25.
[3] *Elementa Physiologiæ*, 1766, viii. p. 245.
[4] Hildebrandt's *Anatomie*, iv. 595, 1832, and Froriep's *Notizen*, October, 1835.

these glands in the cow he believed to be received into the pocket-like depressions of the chorion to which I have previously referred (p. 65), where it comes into relation with the capillary network of the chorion. Von Baer coincided with Weber in regarding the glands as secreting a material to be applied to the nutrition of the ovum. Eschricht looked upon the utricular glands as the sources of the secretion of this nutrient albuminous fluid, whilst he apparently regarded the "cells" (crypts) as the places of formation of ordinary mucus; and this conclusion, at least as regards the function of the glands, has been adopted by various anatomists. Dr Sharpey, from his researches on the bitch, thought it not improbable that in viviparous animals generally, a matter deposited from the maternal system by means of a glandular apparatus may be absorbed into that of the fœtus and serve for its nutrition. Prof. Goodsir in his description of the human placenta[1] showed how the cells of the decidua were prolonged into the interior of the placenta, so as to invest the villi, and regarded them as secreting cells, the remains of the secreting mucous membrane of the uterus; corresponding, therefore, with the cellular lining of the cotyledons of the Ruminants. He considered these cells to perform during intra-uterine life a function similar to that performed after birth by the gastro-intestinal mucous membrane.

Signor Ercolani, of Bologna, whose Memoirs on the Structure of the Placenta in various animals equal in importance and interest the classical Essays of von Baer and Eschricht, has given a more precise aspect to this question. He admits the presence of utricular glands in the mucosa in most orders of mammals, and their increase in size during pregnancy, but conceives that their chief function is to furnish nutritive materials during only the earliest stages of gestation, before the crypts are formed. But further, he points out that in all mammalia, during pregnancy, the surface of the uterine mucosa becomes folded into multitudes of crypts (or follicles as he terms them) in which the fœtal villi are lodged. These follicles form a new glandular organ which prepares a secretion that supersedes that of the utricular glands and is absorbed by the fœtal villi.

---

[1] *Anatomical and Pathological Observations*, pp. 63, 114. 1845.

On this important subject a few words will now be said. There can be no doubt that the Utricular Glands are secreting structures, that they enlarge during pregnancy, and that their secretion is poured out between the mucous membrane and the chorion. In the Diffused form of placenta they have the appearance of being structurally perfect up to the completion of gestation; and their secretion is poured out so as to be brought into direct relation with the inter-villous, and not with the villous portion of the chorion. But as the whole free surface of the chorion is provided with capillaries, the one is no doubt as capable of absorption as the other, and the glands are presumably active throughout intra-uterine life. In the pig, mare, and Lemurs, well-defined areas of vascular chorion are in apposition with the equally well-defined areas on the surface of the mucosa in which the glands opened, and in the Lemurs the mouths of the glands are concentrated in a very remarkable manner in these areas, so that a considerable quantity of secretion would be brought into contact with the apposed non-villous, but vascular, areas of the chorion. In *Orca*, and possibly also to some extent in the Narwhal, the secretion of the glands being poured into some of the crypts is brought into relation with the villi which occupy those crypts; but in other parts of the Narwhal's placenta their secretion is poured out opposite non-villous areas of the chorion. The observations on the Narwhal not only show that the foetus may attain a precise stage of development before any appearance of crypts can be observed, but that after the crypts are fully developed the utricular glands have a diameter much greater than before the crypts had made their appearance. In the Polycotyledonary placenta the utricular glands are not situated in the cotyledons, so that the uterine milk cannot be formed by them. They exist abundantly in the inter-cotyledonary parts of the mucosa, and their secretion is brought in contact with the surface of the chorion in which the curious pockets described by von Baer, Weber, and myself in the cow, and by me in the giraffe, are situated. I agree with Weber in regarding these pockets as receptacles for the secretion, and the vessels in their walls as engaged in its absorption. I may mention that I have seen the smooth surface of the chorion of the sheep smeared with a white substance, apparently the secretion of the utricular glands.

In the Zonary placenta the glands are altered, and degenerated in the placental zone. In the non-placental area of the mucosa they also show an apparent want of structural completeness, at least at the end of gestation. But in the earlier period of intra-uterine life the glands are undoubtedly active, and as their secretion is poured out it is brought into contact with the smooth but very vascular polar portions of the chorion, by which it is in all probability absorbed. I am of opinion therefore that the utricular glands in all these placentæ have a more enduring function in fœtal nutrition than is admitted by Ercolani.

The Crypts, newly formed during pregnancy, possess the structural characters of secreting organs. Each crypt is lined by an epithelium, descended from the epithelial lining of the uterine mucosa; which from the size and appearance of the cells is obviously endowed with great functional activity. This epithelium rests upon a highly vascular, sub-epithelial tissue, the vascularity of which is doubtless proportioned to the amount of secretion formed by the epithelial cells. At one time I was disposed to think that the epithelial lining of the crypts in *Orca* had not a form which one usually associates with the possession of the power of secretion, and that its placenta offered a difficulty in the way of accepting the theory that the crypts were secreting organs. A re-examination, however, of the mucosa in that animal, together with the evidence afforded by the form of the cells lining the crypts in the allied genus, the Narwhal, has satisfied me that the *Cetacea* are no exception to what is seen in other mammals, so that I have now no hesitation in regarding the crypts as secreting organs. The appearance of these crypts in the early stages of placental formation, and their persistence throughout intra-uterine life, though in the zonary form they may become somewhat difficult to recognise, owing to complexities arising during growth, furnish evidence of their importance. The intimate relation which they bear to the villi, which, in the whole series of placentæ described in these Lectures, are lodged within the crypts, shows that the secretion they form is in a position best fitted for being absorbed by the villi.

In no form of placenta has the secretion of the utricular glands been collected and analysed, and indeed the arrangements of parts do not admit of a sufficient amount being collected for that purpose. In the diffused and zonary forms of placentæ the arrangements are not such as to permit the secretion of the crypts to be collected and examined free from mixture with the secretion of the utricular glands. In the polycotyledonary placenta, where the spongy tissue of the maternal cotyledons consists exclusively of crypts, the secretion of uterine milk can not only be shown to be derived from the crypts, but can be obtained in sufficient quantity for analysis. In the analyses published by Prof. Prevost[1], the secretion was stated to contain water, albumen, fibrin, casein, a gelatinous substance, blood-colouring matter, osmazome, fat and salts. Prof. Schlossberger afterwards gave an account[2] of its composition, and stated that it contained water, fat, albumen, salts, no sugar, and that its reaction was acid. More recently Dr Arthur Gamgee has published an analysis[3], in which the aqueous, fatty, albuminous and saline constituents were determined, and the absence both of sugar and casein definitely stated: in his specimens the reaction of the fluid was either alkaline or neuter. From its composition the uterine milk is well suited to act as a nutrient material.

As there are, therefore, two sets of secreting structures in the gravid maternal mucosa, the Glands and the Crypts, each of which has in relation to it a definite and usually distinct surface of the chorion, it may be a matter for consideration how far these secreting organs perform similar or different functions in fœtal nutrition. No definite information can however as yet be given on this matter, which is still open to physiological enquiry. The fact however is not to be doubted that under the stimulus imparted by the presence of the fertilized ovum the uterus undergoes enormous development and growth. The muscular coat increases so as to provide an apparatus capable of expelling by its contraction the fœtus, when its period of intra-uterine development is completed. In the mucous coat the pre-existing utricular glands are enlarged, and in addition

[1] *Ann. des Sc. Nat.*, 1829, xvi. p. 157, and in conjunction with M. Morin in *Mém. de la Soc. de physique de Genève*, 1841, ix. p. 235.
[2] *Ann. der Chemic und Pharm.*, 1855.
[3] *Brit. and For. Medico-Chir. Review*, 1864, xxxiii. p. 180.

multitudes of crypts are developed, in which secretions are produced capable of nourishing the fœtus during its intra-uterine life, so that the maternal placenta is a great secreting organ.

The presence of a ring of colouring matter at the margins of the placenta in the *Carnivora* would seem to show that in the performance of its secreting function certain pigmentary materials may be separated, which perhaps are to be regarded as excretions, just as the bile is separated by the secreting cells of the liver in the performance of its glycogenic function.

If this view of the function of the maternal placenta be admitted, then the current doctrine that the nutrition of the fœtus is provided for by a simple percolation or diffusion of materials through the walls of the vessels from the maternal blood to the fœtal blood can no longer be accepted. Even if the secreting function of the layer of cells on the free surface of the maternal mucosa be disputed by the advocates of the diffusion theory, it would I think be conceded that the presence of such a layer would offer some mechanical obstruction to the direct passage of materials between the two systems of vessels; a difficulty which would be still further increased by the layer of interposed fluid which undoubtedly exists in the cotyledons of the *Ruminantia*.

An important argument in favour of the diffusion theory must not however be forgotten in the discussion of this question. The interesting series of facts collected by Dr Alexander Harvey[1] rendered the conclusion extremely probable that a portion of the blood of the fœtus is constantly passing into the body of the mother. This subject was experimentally investigated by Mr W. S. Savory[2], who found when strychnia was injected into the body of the fœtus, the umbilical cord being intact, that some time afterwards the mother died from symptoms of strychnia poisoning. Now, as we know of no secreting apparatus by which the constituents of the fœtal blood may be separated and transmitted to the mother, it is difficult to see how the poison can have affected the mother except by diffu-

---

[1] On the Fœtus in Utero as inoculating the Maternal with the peculiarities of the Paternal Organism.—*Monthly Journal of Medical Science*, Oct., 1849, Sep., 1850.

[2] An Experimental Inquiry into the effect upon the Mother of poisoning the Fœtus.

sion from the fœtal into the maternal vessels. The passage of the poison does not take place however with great rapidity, as the first tetanic symptoms did not in any of the experiments begin until nine minutes, and in one case even an interval of half an hour elapsed, after the injection of strychnia into the fœtus. But Mr Savory's experiments also show, when into some only of the embryos the poison had been injected, although the mother lived in one instance 19 minutes after she began to be convulsed, the remaining embryos in her womb that had not been injected showed no symptoms of strychnia poisoning. The transmission therefore of the poison from her to them had not occurred either by diffusion, or through the intermediation of a secretion; so that whatever be the mode in which the nutrition of the fœtus be effected the passage of materials from the mother to the fœtus takes place apparently at only a slow rate. At the same time it should be remembered that experiments have been made which prove the passage of materials from the maternal to the fœtal vessels. Thus early in the century A. C. Mayer[1] administered cyanide of potassium to gravid animals and found the salt not only in the umbilical vessels of the fœtus, but in the fluids of the amnion and allantois. M. Flourens also[2] fed a gravid pig with madder, and found not only its own bones, but those of the fœtuses within the womb, coloured. It does not however follow that in these cases the transmission had been due to direct diffusion from one set of vessels to the other, as the saline and colouring substances might have been separated through the action of the secreting cells in the maternal placenta.

But the placenta is also regarded by physiologists as an organ, which not only provides nutriment for the fœtus, but serves as its respiratory apparatus, and it is believed that an interchange of gases takes place between the fœtal and maternal blood-vessels. Undoubtedly there are many facts on record which seem to show that the *fœtus in utero* needs to respire, and that the placenta is the organ in which respiration is conducted. But there is no evidence that the respiratory changes during intra-uterine life are actively carried on. The

[1] *Meckel's Archiv.*, p. 503, 1817.
[2] *Ann. des Sciences Nat.*, 4th series, xii. p. 245.

experiments of Buffon, Legallois and Wm. Edwards[1] indeed show that asphyxia can be resisted by new-born animals for more than half an hour, and that for some days after birth the need of respiration is not so great as at a later period. The interposition of a layer of cells, possessing some thickness, on the surface of the maternal placenta between the two systems of vessels, whether these cells be regarded as secreting or not, necessarily throws a mechanical difficulty in the way of the ready passage of gases from the one set of vessels to the other, and might be used as an argument against the theory of intra-uterine respiration.

It seems to me therefore that, before the problems of nutrition and respiration in the foetus can be regarded as satisfactorily solved, a further series of experiments, with the aid of the additional light which has now been thrown on the structure of the placenta, will require to be made by the physiologist.

---

[1] Quoted in M. H. Milne-Edwards's *Leçons sur la Physiologie*, ii. p. 560.

EXPLANATION OF THE PLATES.

FIGURE 11 was drawn from nature under my superintendence by Mr John R. Reid. Fig. 5 by Mr J. H. Scott, M.B. For the series of microscopic drawings from which the remaining figures have been engraved, I am indebted to my former assistant, Mr J. C. Ewart, M.B.

PLATE I.

Fig. 1. Vertical section through the placenta, *Pl*, of a Cat, about half time (p. 76). *Ch*, the chorion, the vessels of which are coloured blue, so that the blue network which passes through the thickness of the placenta represents the vessels of the villi; *D*, the decidua serotina; the red-coloured vessels are the vessels of the uterine mucosa, which ascend in the walls of the crypts as far as the chorion, where they not unfrequently show considerable dilatations, *s*; *b*, bud-like terminal offshoot of a villus penetrating into the serotina; *ms*, the muscular coat. × Hartnack 3 obj. 3 Oc.

Fig. 2. Horizontal section through the placenta of the same Cat. *VV*, transversely divided sinuous villi, the capillaries in which are coloured blue; *l, l*, laminæ of uterine mucosa, forming the walls of the crypts. The red colour represents maternal vessels, as in *Fig.* 1.

Fig. 3. Villi, *V*, with a portion of the chorion, *ch*, of the shed placenta of a Cat at full time; *b*, a terminal bud, such as in *Fig.* 1 *b* penetrates deeply into the serotina. × 40.

Fig. 4. Horizontal section through the placenta of a Fox (p. 85). *V*, the blue-coloured vessels of the fœtal villi; *m*, the transversely divided colossal maternal capillaries.

Fig. 5. Section through the placenta of a Fox magnified to show the secreting epithelium. *mm*, the colossal maternal capillaries injected red. *e, e*, the columnar epithelial cells which invest the maternal vessels. *vc*, the vessels of the fœtal villi injected blue. × 160.

Fig. 6. Vertical section through the non-gravid uterine mucosa of a Bitch (p. 84). *e*, ends of columnar epithelial cells covering free surface of mucosa; *g*, tubular gland shown in its entire length; *g'*, a tubular gland cut short. At *a* the continuity of an apparently short gland, with the deeper end of a tube, is shown. *ct*, interglandular connective tissue, with its corpuscles; *b*, arteries passing into the mucosa; *ms*, muscular stratum × 100.

Fig. 7. Horizontal section through the non-gravid uterine mucosa of a Bitch, near the free surface. The close relation which the glands have to each other is shown; also the small proportion of interglandular tissue, in which rounded cells, not unlike lymph or white blood-corpuscles, may be seen. × 100.

Fig. 8. Surface view of the non-gravid uterine mucosa of the Crested Seal (*Cystophora cristata*). At *e* the columnar epithelium is *in situ*, elsewhere it has been removed; *g*, mouth of a tubular gland. The capillary network of the mucosa is coloured red. × 100.

PLATE II.

Fig. 9. Vertical section through the placental area of the mucosa of the Cat, described on p. 72. *cr*, the layer of branching crypts; the epithelial lining of the crypts *ep, ep*, and their highly corpusculated connective-tissue walls *ct, ct*, are represented; *gl*, the glandular layer; the glands are seen in section, much less numerous than the crypts, and surrounded by connective tissue; *ms*, the muscular coat. × Hartnack 3 obj. 4 Oc.; tube out.

Fig. 10. Horizontal section through the crypt-layer of the same uterus. *cr, cr,* cavities of the crypts with their epithelial lining. At *e* the epithelium covering the free surface of the walls of the crypts is seen; at *e'* the walls are in section, and the sub-epithelial connective tissue, with its corpuscles, *ct,* is exposed. × Hartnack 7 obj. 3 Oc.; tube out.

Fig. 11. Portion of placenta, *Pl,* of the Grey Seal partially dissected off the uterus. *mm,* the uterine mucosa forming the non-deciduous part of the serotina; folds of this membrane may be seen entering the sulci or primary fissures, *ss,* between the convolutions of the placenta. The broader red lines on the exposed surface of the placenta are intended to represent the secondary fissures of the placenta, and the finer lines the tertiary fissures, by which it is subdivided into the ultimate lobules, *l. np, np,* non-placental portions of the mucous membrane. Natural size.

Fig. 12. Surface view of the uterine mucosa of the same Seal forming the non-deciduous serotina. At *e, e* the broad ends of the columnar epithelium cells, still *in situ,* are represented. In the rest of the figure the epithelium has been removed. *uv,* larger trunks of the blood-vessels of the mucosa; *c,* capillary network; *ct,* corpusculated sub-epithelial connective tissue; *gl,* portion of one of the utricular glands. × 250.

PLATE III.

Fig. 13. Vertical section through one lobule and a portion of an adjacent lobule of the placenta of the Grey Seal. *Ch,* chorion; *VV,* stems of the large arborescent villi; *V',* smaller villi arising directly from the chorion. The blue-coloured vessels in the chorion and villi are the ramifications of the umbilical vein. *g.g.,* greyish membrane at the periphery of the lobule in which the blue-coloured vessels of the villi ramify; *b.b.,* bud-like offshoots of the finer branches of the villi; *mm,* uterine mucosa forming the non-deciduous serotina in relation with the placental lobule; *gl,* utricular gland; *uv,* uterine blood-vessels, coloured red, passing into *f,* a tertiary fissure between the two lobules. At the upper end of this fissure these vessels form a network continuous with the intra-lobular maternal capillaries. At *tr* the intra-placental trabecular arrangement of the mucosa is shown isolated and drawn away from the finer branches of the villi. In the greater part of this figure the maternal trabeculæ are shown *in situ* intertwined amidst the fœtal villi. × 40.

Fig. 14. Uterine surface of four of the lobules of the placenta. *g.g.,* greyish membrane forming the periphery of the lobules; at *g'* the membrane has been dissected off; the blue-coloured vessels are branches of the umbilical vein; *t,* processes of the uterine mucosa, with the vessels coloured red, dipping into the tertiary fissures between the lobules. × 4.

Fig. 15. Intra-placental maternal trabeculæ, *tr.* At *ep, ep* the columnar secreting epithelial covering is seen *in situ;* at *ep', ep'* partially shed. Elsewhere the epithelium has been entirely removed, so as to show the sub-epithelial tissue with its corpuscles, and the single capillary *c,* in each trabecula. *V,* a single bud-like offshoot of a villus lying between the trabeculæ in contact with their epithelium. × 200.

Fig. 16. Branch of a villus, *V,* with its terminal bud-like offshoots. The vascularity of the villus is shown in the upper part of the figure, whilst at the lower end its cellular structure is represented. × 200.

CAMBRIDGE: PRINTED BY C. J. CLAY, M.A. AT THE UNIVERSITY PRESS.

www.ingramcontent.com/pod-product-compliance
Lightning Source LLC
Chambersburg PA
CBHW021938190326
41519CB00009B/1063